Unified Chromatography

ACS SYMPOSIUM SERIES **748**

Unified Chromatography

J. F. Parcher, EDITOR
University of Mississippi

T. L. Chester, EDITOR
The Procter and Gamble Company

American Chemical Society, Washington, DC

Library of Congress Cataloging-in-Publication Data

United Chromatography / J.F. Parcher, editor, T.L. Chester, editor

    p.   cm.—(ACS symposium series, ISSN 0097–6156 ; 748)

    Includes bibliographical references and index.

    ISBN 0–8412–3638–0

    1. Chromatography—Congresses.

    I. Parcher, J.F., 1940– . II. Chester, T.L., 1949– . III. Series.

QD79.C4 U55   1999
543´.089—dc21   99–47736

The paper used in this publication meets the minimum requirements of American National Standard for Information Sciences—Permanence of Paper for Printed Library Materials, ANSI Z39.48–1984.

Copyright © 2000 American Chemical Society

Distributed by Oxford University Press

All Rights Reserved. Reprographic copying beyond that permitted by Sections 107 or 108 of the U.S. Copyright Act is allowed for internal use only, provided that a per-chapter fee of $20.00 plus $0.25 per page is paid to the Copyright Clearance Center, Inc., 222 Rosewood Drive, Danvers, MA 01923, USA. Republication or reproduction for sale of pages in this book is permitted only under license from ACS. Direct these and other permissions requests to ACS Copyright Office, Publications Division, 1155 16th Street, N.W., Washington, DC 20036.

The citation of trade names and/or names of manufacturers in this publication is not to be construed as an endorsement or as approval by ACS of the commercial products or services referenced herein; nor should the mere reference herein to any drawing, specification, chemical process, or other data be regarded as a license or as a conveyance of any right or permission to the holder, reader, or any other person or corporation, to manufacture, reproduce, use, or sell any patented invention or copyrighted work that may in any way be related thereto. Registered names, trademarks, etc., used in this publication, even without specific indication thereof, are not to be considered unprotected by law.

PRINTED IN THE UNITED STATES OF AMERICA

Advisory Board

ACS Symposium Series

Mary E. Castellion
ChemEdit Company

Arthur B. Ellis
University of Wisconsin at Madison

Jeffrey S. Gaffney
Argonne National Laboratory

Gunda I. Georg
University of Kansas

Lawrence P. Klemann
Nabisco Foods Group

Richard N. Loeppky
University of Missouri

Cynthia A. Maryanoff
R. W. Johnson Pharmaceutical
    Research Institute

Roger A. Minear
University of Illinois
    at Urbana–Champaign

Omkaram Nalamasu
AT&T Bell Laboratories

Kinam Park
Purdue University

Katherine R. Porter
Duke University

Douglas A. Smith
The DAS Group, Inc.

Martin R. Tant
Eastman Chemical Co.

Michael D. Taylor
Parke-Davis Pharmaceutical
    Research

Leroy B. Townsend
University of Michigan

William C. Walker
DuPont Company

Foreword

THE ACS SYMPOSIUM SERIES was first published in 1974 to provide a mechanism for publishing symposia quickly in book form. The purpose of the series is to publish timely, comprehensive books developed from ACS sponsored symposia based on current scientific research. Occasionally, books are developed from symposia sponsored by other organizations when the topic is of keen interest to the chemistry audience.

Before agreeing to publish a book, the proposed table of contents is reviewed for appropriate and comprehensive coverage and for interest to the audience. Some papers may be excluded in order to better focus the book; others may be added to provide comprehensiveness. When appropriate, overview or introductory chapters are added. Drafts of chapters are peer-reviewed prior to final acceptance or rejection, and manuscripts are prepared in camera-ready format.

As a rule, only original research papers and original review papers are included in the volumes. Verbatim reproductions of previously published papers are not accepted.

ACS BOOKS DEPARTMENT

Contents

Preface ix

1. Unified Chromatography: What Is It? 1
   J. F. Parcher and T. L. Chester

THEORETICAL AND GENERAL CONSIDERATIONS

2. Unified Chromatography from the Mobile Phase Perspective 6
   T. L. Chester

3. Pressure as a Unifying Parameter in Chromatographic Retention 30
   C. E. Evans, M. C. Ringo, and L. M. Ponton

4. Three-Dimensional Stochastic Simulation for the Unified Treatment
   of Chromatographic and Electrophoretic Separations 37
   Victoria L. McGuffin, Peter E. Krouskop, and Daniel L. Hopkins

5. Computer Simulations of Interphases and Solute Transfer
   in Liquid and Size Exclusion Chromatography 67
   Thomas L. Beck and Steven J. Klatte

6. Exploring Multicomponent Phase Equilibria
   by Monte Carlo Simulations: Toward a Description
   of Gas–Liquid Chromatography 82
   Marcus G. Martin, J. Ilja Siepmann, and Mark R. Schure

7. Stationary- and Mobile-Phase Interactions in Supercritical
   Fluid Chromatography 96
   Sihua Xu, Phillip S. Wells, Tingmei Tao, Kwang S. Yun,
   and J. F. Parcher

TECHNIQUES AND SPECIFIC APPLICATIONS

8. Packed Capillary Columns in Hot Liquid and in Supercritical
   Mobile Phases 120
   T. Greibrokk, E. Lundanes, R. Trones, P. Molander, L. Roed,
   I. L. Skuland, T. Andersen, I. Bruheim, and B. Jachwitz

9. **Packed Capillary Column Chromatography with Gas, Supercritical, and Liquid Mobile Phases** 142
    Keith D. Bartle, Anthony A. Clifford, Peter Myers,
    Mark M. Robson, Katherine Seale, Daixin Tong,
    David N. Batchelder, and Suzanne Cooper

10. **Applications of Enhanced-Fluidity Liquid Mixtures in Separation Science: An Update** 168
    S. V. Olesik

11. **Universal Chromatography for Fast Separations** 179
    Milton L. Lee and Christopher R. Bowerbank

12. **Practical Advantages of Packed Column Supercritical Fluid Chromatography in Supporting Combinatorial Chemistry** 203
    T. A. Berger

INDEXES

Author Index 236

Subject Index 237

Preface

We became interested in the concept of unified chromatography as we considered the evolution of chromatographic theory and the recent investigations of several researchers pushing chromatography out of the norms of conventional parameter space. What happens, we wondered, if researchers use various conventional techniques as their starting points and push the boundaries of these techniques toward each other? Will researchers find a true boundary separating the various practices they already know, or will the boundaries merge? If so, will one theory suffice to explain all of chromatography regardless of the fluid state of the mobile phase? Are there more general treatments of the role of temperature, pressure, surface effects, molecular interactions, and so on that might bridge (or separate) the conventional techniques? What ramifications would all this have on the evolution of chromatographic practice and instrumentation? What effect, if any, would it have on the manner in which critical parameters are chosen in our never-ending search for optimal separations?

To address this issue we organized a symposium of researchers doing some unusual things with chromatography. We invited speakers working with high temperatures, high pressures, unusual fluids, and unconventional conditions, as well as others looking at intermolecular interactions and other theories. This book is a compilation of some of those presentations plus a few postsymposium additions. The symposium was held at the American Chemical Society (ACS) National Meeting in Boston, Massachusetts, in August 1998, and was sponsored by the ACS Division of Analytical Chemistry. We hope that you find both the subjects and the perspectives stimulating.

Each of the papers in this volume was peer-reviewed by individuals who freely contributed their time and effort to ensure the quality of the final product. We particularly wish to thank these reviewerws as well as the authors for all their efforts. We also thank the members of the Division of Analytical Chemistry Executive Committee, and especially Henry N. Blount III, for their encouragement and support throughout this project. Finally, we wish to acknowledge the generous contributions of the Rohm and Haas Company and the Procter and Gamble Company for providing financial support for the original symposium.

J. F. PARCHER
Chemistry Department
University of Mississippi
University, MS 38677
chjfp@olemiss.edu

T. L. CHESTER
The Procter and Gamble Company
Miami Valley Laboratories
P.O. Box 538707
Cincinnati, OH 45253–8707
chester.tl@pg.com

Chapter 1

Unified Chromatography: What Is It?

J. F. Parcher[1] and T. L. Chester[2]

[1]Chemistry Department, University of Mississippi, University, MS 38677
[2]The Procter and Gamble Company, Miami Valley Laboratories,
P.O. Box 538707, Cincinnati, OH 45253-8707

In this chapter we explain why we are interested in this concept, give some of the history, introduce some of the topics to be treated in the subsequent chapters, and set the stage for much further consideration.

Today we are at an extremely exciting point in the development and refinement of chromatography. Researchers are recognizing that some of the perceived limits of the past are not real. We are using gases as mobile phases but compressing them until they have liquid-like properties. We are condensing what are ordinarily gases into liquids and using them as mobile phases at subambient temperatures where ordinary liquids are frozen solid. We are using ordinary liquids above their normal boiling points but still keeping them in the liquid state by the application of pressure. We are finding ways of reducing mobile phase viscosity, increasing diffusion rates, increasing analysis speed, tuning selectivity, and explaining retention more comprehensively. And we are removing the barriers that artificially distinguish and limit essentially similar chromatography techniques, thus opening new understanding and even more new possibilities. This is all included in the concept we call *unified chromatography*.

We are not sure how far the concept can be developed or what additional benefits may arise. In fact, the only thing we are really sure of is what unified chromatography is not. The antithesis of the concept of unified chromatography is the often accepted practice that the methods, instrumentation, and theories employed in any particular separation technique, for example gas chromatography, are completely unrelated to those used in other separation techniques like supercritical fluid chromatography, liquid chromatography, or electrophoresis.

We organized the symposium, from which this book was derived, to bring together people working with different aspects of chromatography. We wanted these people to consider the underlying similarities of their work, test the strength of the traditional boundaries separating chromatography into individual techniques, and help catalyze even more new work and understanding.

The concept of "One World of Chromatography" has been an intriguing, albeit somewhat ethereal, Holy Grail for chromatographers for decades. Different individuals have proposed the idea of a unified theory, others a single instrument, while yet another group has advocated a single technique that would encompass all of the technology currently available.

Martire has been a major proponent of a unified approach to the theory of chromatography. He first used the words *unified theory* in the titles of a set of papers (*1,2*) published in the early 1980s. The models were based on a statistical mechanical treatment of the equilibrium distribution of a solute between the stationary and mobile phases in liquid chromatography. While the initial work involved only liquid chromatography, the effort was soon extended to include liquid, gas, and supercritical fluid chromatography (*3-6*). The most recently developed models are based on the lattice fluid model first proposed by Sanchez and Lacombe in 1978 (*7,8*) and subsequently enhanced by Sanchez and his colleagues (*9-12*). The lattice fluid models are complex, but currently represent the only extant theory that can be used to interpret, let alone predict, the retention volume of a chromatographic solute as a function of pressure, temperature, and mobile phase composition and/or density.

Analytical and physical chemists have long been using experimental results from phase distribution measurements to develop theoretical models to interpret retention volume data from chromatographic experiments. Chemical engineers have, in a similar manner, often used chromatographic results to describe the solubility of solids in supercritical fluids or mixtures of supercritical fluids and liquids in polymers. Shim and Johnston (*13-15*) studied the partition of toluene between fluid and polymeric phases. More recently, Eckert *et al.* (*16*) studied the effect of various modifiers, *viz.,* methanol, acetone, and isopropanol, on the swelling of poly(dimethylsiloxane) caused by the absorption of modifier by the polymer over a range of temperatures and pressures. Eckert *et al.* (*17*) used the Sanchez-Lacombe lattice fluid model, the same basic idea used by Martire for his Unified Theory of Chromatography, to model the swelling, absorption, and isothermal phase behavior for the modifiers. Again, the lattice fluid models are complex, but their ability to describe such complex systems involving a solute, supercritical CO_2 and a polar liquid modifier, all in equilibrium with a polymeric stationary liquid phase, is quite remarkable and bodes well for our ability to develop a truly comprehensive theory for a hypothetical unified chromatography. The collaboration between chemists and engineers is encouraging and most certainly necessary for our further understanding of the fundamental mechanisms that control all types of separations.

There also has been a push to develop instrumentation that could be used in a unified approach to chromatography. Ishii and Takeuchi (*18,19*) used the titles *Unified Fluid Chromatography* and *Unified Capillary Chromatography* to describe their work on the development of a single chromatographic instrument that could perform GC-, SFC-, or HPLC-like separations depending upon the temperature and pressure of the mobile phase. Recently, many individuals and groups have espoused the same idea, but to our knowledge, Ishii's group was the earliest to use the word *unified* in the title of a published article describing multi-use chromatographic instrumentation.

In 1991, the late Professor J. C. Giddings published a well-received book entitled *Unified Separation Science* (*20*). The volume was meant to be a textbook for a graduate-level course in separations. However, the author took advantage of the privilege of authority to remind those of us who teach such courses or workshops that "...it is more important to understand how it works than how to do it; basics rather than recipes. Procedures will change and evolve and old recipes are soon obsolete, but the underlying mechanisms of separation will be around for a long time."

More recently, several authors have addressed the idea of unified chromatography rather than a particular implementation of the concept. In particular, Chester (*21,22*) has discussed what he calls "Chromatography from the Mobile Phase Perspective". He emphasizes the measurement and interpretation of phase diagrams of mobile phase fluids as a means of defining the chromatography and eliminating unnecessary boundaries between named techniques. The relation between the phase behavior of mobile phases and the resultant chromatographic performance of chromatographic systems is a fundamental relationship that is often overlooked. The primary reasons for such neglect are the experimental difficulty involved in the determination of phase diagrams, especially when binary or ternary fluids are considered, and the somewhat arcane nature of phase diagrams in general.

We have based these introductory comments on a few of the chromatography publications where the term *unified* was mentioned; however, it is obvious that myriad publications could be cited in an effort to eliminate the artificial barriers established historically between the various forms of chromatography or separation schemes in general. Most of the so-called two-dimensional chromatography schemes can be viewed as progress toward a single, combined form of chromatography. And, finally, almost all of the papers presented at the Symposium, as well as the chapters appearing in this volume, present cross-cutting, state-of-the-art research by some of the old-timers and new-comers in the field of separation science. If these presentations and articles help to eliminate some of the prejudice and tunnel vision evident in the chromatographic literature, then we will have fulfilled a portion of our original goal in opening for discussion the alluring concept of *unified chromatography*.

Literature Cited

1. Boehm, R. E.; Martire, D. E. *J. Phys. Chem.* **1980,** *84*, 3620-3630.
2. Martire, D. E.; Boehm, R. E. *J. Phys. Chem.* **1983,** *87*, 1045-1062.
3. Martire, D. E. *J. Liq. Chromatogr.* **1988,** *11*, 1779-1807.
4. Martire, D. E.; Boehm, R. E. *J. Phys. Chem.* **1987,** *91*, 2433-2446.
5. Martire, D. E. *J. Liq. Chromatogr.* **1987,** *10*, 1569-1588.
6. Martire, D. E. *J. Chromatogr.* **1988,** *452*, 17-30.
7. Sanchez, I. C.; Lacombe, R. H. *J. Phys. Chem.* **1976,** *80*, 2352-2362.
8. Sanchez, I. C.; Lacombe, R. H. *Macromolecules* **1978,** *11*, 1145-1156.
9. Sanchez, I. C. *Encyclopedia of Physical Science & Technology* **1986,** *11*, 1-18.

10. Sanchez, I. C.; Rodgers, P. A. *Pure & Appl. Chem.* **1990**, *62*, 2107-2114.
11. Sanchez, I. C. *Macromolecules* **1991**, *24*, 908-916.
12. Sanchez, I. C.; Panayiotou, C. G. In *Models for Thermodynamics and Phase Equilibrium Calculation;* Sandler, S., Ed., Marcel Dekker: New York, 1994; pp187-246.
13. Shim, J.-J.; Johnston, K. P. *A.I.Ch.E., J.* **1989**, *35*, 1097-1106.
14. Shim, J.-J.; Johnston, K. P. *A.I.Ch.E., J.* **1991**, *37*, 607-616.
15. Shim, J.-J.; Johnston, K. P. *J. Phys. Chem.* **1991**, *95*, 353-360.
16. Vincent, M. F.; Kazarian, S. G.; West, B. L.; Berkner, J. A.; Bright, F. V.; Liotta, C. L.; Eckert, C. A. *J. Phys. Chem. B* **1998**, *102*, 2176-2186.
17. West, B. L.; Bush, D.; Brantley, N. H.; Vincent, M. F.; Kazarian, S. G.; Eckert, C. A. *Ind. Eng. Chem. Res.* **1998**, *37*, 3305-3311.
18. Ishii, D.; Toyohide, T. *J. Chromatogr. Sci.* **1989**, *27*, 71-74.
19. Ishii, D.; Niwa, T.; Takeuchi, T. *JHRCC & CC* **1988**, *11*, 800-801.
20. Giddings, J. C. *Unified Separation Science*; John Wiley & Sons, Inc, 1991.
21. Chester, T. L. *Anal. Chem.* **1997**, *69*, 165A-169A.
22. Chester, T. L. *Microchem. J.* **1999**, *61*, 12-24.

Theoretical and General Considerations

Chapter 2

Unified Chromatography from the Mobile Phase Perspective

T. L. Chester

The Procter and Gamble Company, Miami Valley Laboratories, P.O. Box 538707, Cincinnati, OH 45253-8707

Liquid, Subcritical Fluid, Enhanced Fluidity, Supercritical Fluid, Hyperbaric, Solvating Gas, and Gas Chromatographies are merged into a continuous model with no behavior gaps and no discontinuities. The distinctions between individually named techniques vanish in this unified model. It is based on the general phase behavior of Type I binary mixtures (in which the two components are miscible as liquids). Conventional Gas Chromatography and Liquid Chromatography are simply limiting cases of the unified model and exist when the column-outlet pressure is ambient. Seeking the optimal conditions within the allowed parameter space leads to significant separation improvements. A unified chromatograph, with only a few new capabilities not already found on a conventional liquid chromatograph, already exists commercially and can perform all of the named techniques except conventional GC.

The characteristics and outcome of a chromatographic separation not only depend on our selection of the stationary phase, mobile phase, phase ratio, velocity, etc., but also, more frequently than we may realize, on environmental factors. These are easily and often overlooked, especially in the practice of liquid chromatography (LC) or high-performance LC (HPLC).

LC has been practiced throughout most of its history at ambient temperature and outlet pressure. Column temperature control is common today, and temperatures somewhat above ambient are often chosen to adjust the selectivity, lower the mobile-phase viscosity, and improve diffusion rates (*1-3*). But the full exploitation of temperature as an LC parameter is extremely limited when the column outlet pressure is left to default to one atmosphere (approximately 0.1 MPa)--LC cannot be performed at temperatures above the mobile-phase boiling temperature corresponding to the column outlet pressure (*3-5*).

The developers of gas chromatography (GC) recognized very early that ambient temperature was not the best choice for the column. GC is quite sophisticated today, compared to its early history, in its use of temperature control to enhance injection, solute focusing, separation, and mass transfer to detectors. In contrast, relatively little attention has been paid to mobile-phase selection and outlet pressure in GC, or to the effect outlet pressure would have on mass-transfer properties when non-inert mobile phases are considered. So far, this has been predominantly the realm of researchers developing supercritical fluid chromatography (SFC) and related techniques (*6-11*).

This chapter will show how simply controlling what were originally environmental parameters--the column temperature *and* the column outlet pressure--opens a new dimension of control and performance in chromatography and unifies all the column chromatography niche techniques (including LC and GC) into one simple and complete picture (*12*).

The Deceptions of Ambient Conditions

Surely life on Earth would be much different, or would not exist at all, if there were no liquid water present. Most life can only exist in a very narrow temperature range as compared to the extremes that exist terrestrially. The comfort range for humans is even smaller. Ambient pressure is almost as important to life as is temperature since liquid water can only exist in equilibrium with its vapor when the atmospheric pressure exceeds the vapor pressure of water. We have been blessed by Nature with ambient conditions on Earth that are, at the very least, very convenient to us.

These same ambient conditions are also a curse and an impediment to fuller understanding of the true nature of things. Constant immersion in an "ambient" atmosphere that is 295 °K (give or take a few degrees), with a pressure never varying much from 0.1 MPa, composed of 21% oxygen, 78% nitrogen, etc., creates many experiences for us that we take for granted. Just imagine how long it took in the history of humankind to develop today's concept of air. Because of our constant exposure to environmental norms, the dependence of many behaviors on these norms is difficult to fully realize.

Take, for example, our concept of fire. The National Fire Protection Association (of the U. S. A.) states in defining their fire triangle, "In order to have a fire, there must be three elements: fuel, heat, and air (more specifically, oxygen)" (*13*). However, this model is only valid in places where the "air" contains sufficient oxygen to support oxidative combustion. Imagine how different our explanation of fire, or of fuel for that matter, would be if our atmosphere were hydrogen, methane, or ammonia (assuming we were still here, of course). The point is that many or our expectations, and even many of our definitions, result from our ambient environmental conditions.

What makes a solvent good for LC? We judge fluids as useful for particular purposes based on our experience in dealing with them at or quite close to ambient conditions. The first consideration in choosing a material for use as an LC solvent is its liquid behavior. Obviously, anything not liquid at ambient conditions is rejected immediately. But in addition, "good" liquids would not be too volatile to use conveniently, nor too viscous to pour or pump. Only materials passing these requirements are then considered further for their ability to dissolve and transport solutes. One "good" fluid for our typical purposes is hexane. We would seldom consider using decane as an LC solvent since hexane has similar polarity, is still not too volatile at laboratory temperatures, and is much easier to pump through packed columns than decane.

Similarly, we would seldom (if ever) think of using butane as a solvent in LC. It would be very inconvenient to store according to our usual LC expectations, and would require equipment modifications for pumping since it is a gas at ambient conditions. In pressure-driven LC systems, butane would exist in a liquid state at the inlet pressure of most columns (even if the temperature were elevated a few degrees above ambient), but would boil somewhere along the column length as the pressure diminishes upon nearing the column outlet. The outlet pressure, conveniently, is held quite close to 0.1 MPa by our atmosphere. (If the outlet tubing, detector, fraction collector, or any other equipment downstream from the column have any resistance to flow, the column outlet pressure can be elevated perhaps to as much as 3.5 MPa. Most practitioners consider such "high" outlet pressure undesirable since it can damage flow-through detector cells. This pressure is seldom monitored and almost never controlled.) Clearly, our choices for appropriate solvents, and our goodness measures of LC performance, are highly influenced by our ambient conditions.

Now imagine what HPLC would be like if it had been developed in parallel with us at another location where the ambient conditions may be just a bit colder and the pressure a little higher--perhaps a place like Jupiter. How strongly would a stubborn Jovian chromatographer defend his or her (or its) ambient conditions as the correct ones and reject any notion to try different conditions? As strongly as we defend our ambient conditions, most likely! And what would be a good non-polar solvent on Jupiter? Jupiter is much too cold even for using butane.

So, who would be correct? The road to Unified Chromatography begins with the realization that our ambient conditions may not be best for every purpose, and that no single set of conditions, including the column temperature and outlet pressure, is necessarily best for every separation.

Phase Diagrams to Describe the Chromatographic Mobile Phase--Pure Fluids

To continue down this road to unified chromatography it is necessary to consider the phase behavior of fluids that we use as mobile phases. A typical phase diagram for a pure substance is a pressure (P)-temperature (T) plot as shown in Figure 1. A boundary, the boiling line, separates the liquid (l) and vapor (v) states. Parameters like the density, dielectric constant, viscosity, etc. change continuously throughout both the liquid and vapor states, but when the boiling line is crossed and a liquid is converted to a gas by boiling (or vice versa by condensation) the fluid properties are changed discontinuously.

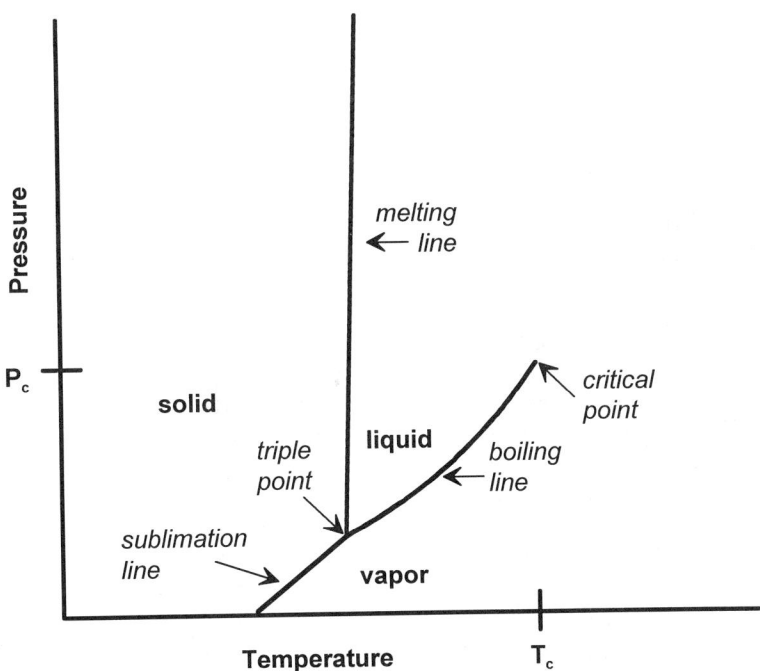

Figure 1. Pressure-temperature phase diagram for a pure substance. The distinction between the liquid and vapor phases ceases to exist at the critical point.

When conditions are adjusted to any point on the boiling line, liquid and vapor can co-exist in equilibrium. Of course, the liquid and vapor will have very different properties. If we then raise the temperature and pressure in concert to keep the two fluids co-existing in equilibrium while moving along the boiling line to higher temperatures and pressures, we would observe the corresponding properties in the two separate phases approaching each other. For example, the density of the liquid (which starts considerably higher than that of the vapor) will decrease, and that of the vapor will increase. Thus, the densities of the two fluids approach each other as the temperature and pressure are increased along the boiling line.

If we continue traveling on the boiling line to higher temperatures and pressures, the two separate fluid phases continue becoming more and more alike. Eventually we will reach a point where the liquid and vapor phases become indistinguishable. The critical point is the point on the boiling line where the liquid and vapor phases merge. The coordinates of this point are defined as the critical temperature (T_c) and the critical pressure (P_c). At and beyond the critical point there are no longer separate liquid and vapor phases, but just one fluid phase which shares the properties of liquid and vapor.

The region of the phase diagram at temperatures and pressures higher than the critical temperature and pressure values is formally (and arbitrarily) designated as the *supercritical fluid* region by both the American Society for Testing and Materials (ASTM) and by the International Union of Pure and Applied Chemistry (IUPAC) (*14,15*). This is indicated in Figure 2. This unfortunate designation introduces what *appears* to be a fourth state of matter, the supercritical fluid. This is an *immense* source of confusion among novices and even some experts. The literature is full of statements regarding the *transition* between a liquid and a supercritical fluid phase, or between a vapor and a supercritical fluid phase. *This is incorrect.* Discontinuous phase change occurs when the boiling line is crossed, but no discontinuous transitions or phase changes take place for isothermal pressure changes above the critical temperature or for isobaric temperature changes above the critical pressure. *There are no transitions into or out of a supercritical fluid state even though the supercritical fluid region is defined formally.* The distinction between liquid and vapor simply ceases for temperatures and pressures beyond the critical point. Figure 1, not Figure 2, is the accurate depiction of phase behavior.

Instead of the familiar notion of discontinuous liquid and vapor states separated by boiling, as reinforced by our experience living in our water-rich world at more-or-less constant pressure, a more useful depiction of fluids for chromatographers is shown in Figure 3. Here we see a continuous one-phase region encompassing the ordinary liquid and vapor states. Ordinary liquids and vapors as we know them are, in essence, limits of this continuum. Note that it is possible to convert a liquid to a vapor, or a vapor to a liquid, without undergoing a discontinuous phase transition by choosing a pressure-temperature path that is wholly within the continuum. The required path simply goes around the critical point and avoids going through the

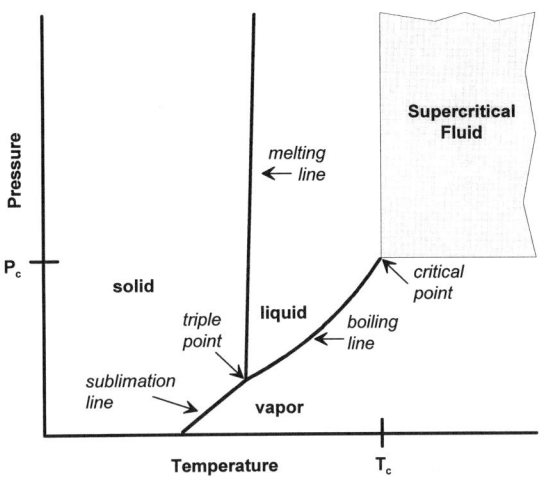

Figure 2. The supercritical fluid region is formally defined as shown, however the apparent boundaries are *not* phase transitions, only arbitrary definitions. Figure 1 is the more accurate representation of the actual phase behavior.

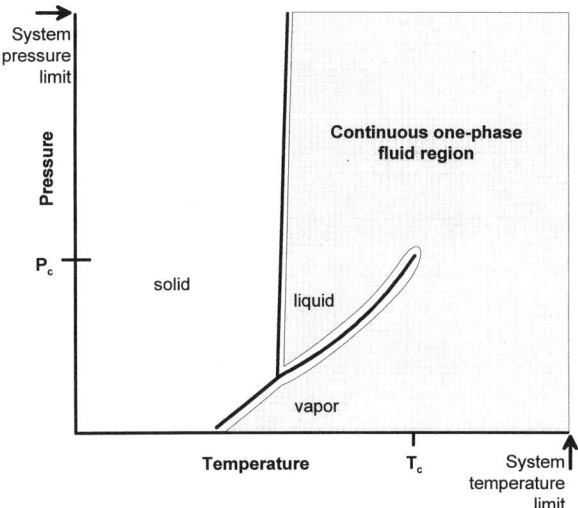

Figure 3. The fluid continuum available to chromatographers.

boiling line. Also note that the maximum temperature and pressure we can safely use with our system also place limits on our use of the continuum.

It is usually considered necessary to avoid phase transitions on a chromatographic column. Thus, we have to avoid crossing the boiling line as conditions (particularly the pressure) change between the column inlet and outlet. Thus, the temperatures and pressures near the boiling line and critical point need to be placed off limits to us over the entire column length. However, we are completely free to select conditions anywhere else within the fluid continuum shown in Figure 3 and take full advantage of whatever "unusual" mobile-phase properties we find.

In the liquid region well below the critical temperature, the mobile phase is nearly incompressible. Therefore, in a typical, conventional HPLC experiment, varying the pressure has little effect on the mobile-phase strength. However, as the temperature is increased (perhaps with a suitable pressure increase to prevent boiling), the mobile phase becomes less viscous, diffusion rates increase, the liquid becomes increasingly compressible, and its strength becomes pressure-dependent. At temperatures and pressures just above the critical point, the fluid is highly compressible and pressure has a very large influence on the mobile-phase strength.

For temperatures above T_c the solvent strength is continuously adjustable by means of pressure adjustment. The fluid has liquid-like solvent strength when compressed to liquid-like densities. The solvent strength diminishes as the density is reduced until, at sufficiently low pressures, the fluid behavior approaches that of a perfect gas with no significant intermolecular forces and no solvent strength. This gas behavior, of course, extends continuously to lower temperatures when the pressure is low enough. Thus, we have the ability to continuously tune the solvent strength from zero to liquid-like values by varying the temperature and pressure.

Binary Mixtures. In chromatography we are often interested in using a multi-component mobile phase instead of a pure fluid, particularly when we think of doing LC. It is useful to continue thinking of the fluid phase behavior, but this requires us to expand the phase diagram to include the composition variation possible in a binary mobile phase. We will only consider binary fluids here, but the general principles also apply to more complicated systems.

Six general types of binary-mixture systems have been defined (*16*). Some of these systems have large miscibility gaps rendering them useless for chromatography over much of their composition ranges. However, Type I mixtures are the simplest and most widely used mixtures in LC. These are the mixtures in which the two components are miscible in all proportions as liquids.

To consider the phase behavior of a binary mixture it is necessary to add to the phase diagram a third axis representing the fluid composition. We will build a Type I binary phase diagram from the phase behavior of the two, pure, fluid components. We will simply call the two components *a* and *b*. The choices of *a* and *b* are completely arbitrary in this exercise except that *a* will be used to designate the more volatile component, and we will restrict the *a* and *b* choices to materials that together form a Type I binary mixture.

What we see in Figure 4 are the boiling lines (shown as solid lines) for pure *a* and pure *b* as they exist in the planes at the limiting values of the composition axis for their binary mixture. These limiting planes correspond to the single-component phase diagrams like that in Figure 1. The triple points and the solid regions have been omitted from Figure 4 for clarity of the liquid-vapor (*l-v*) behavior. The boiling lines for *a* and *b* end at their critical points. Every intermediate mixture composition between pure *a* and pure *b* also has a corresponding critical point. The dashed line in Figure 4 depicts the locus of these mixture critical points spanning the composition dimension and connecting the critical points for *a* and *b*.

We have seen that a pure material co-exists in the liquid and vapor states at equilibrium on a line (that is, the boiling line) in a two-dimensional phase diagram (as in Figure 1). However, the two-phase *l-v* region for a Type I binary mixture is a volume in three dimensions (pressure, temperature, and composition) as shown by the shaded section of Figure 5. Note that the critical locus runs over the top of the two-phase region. This region collapses into the boiling lines for pure *a* and pure *b* at the limits of the composition axis at temperatures below critical for each pure fluid. This is difficult to visualize in print, and is further complicated in this case because the shaded section representing the two-phase region has been chopped off at 25 °C to show the shape of the isotherm at that temperature. Much of the *a* boiling line visible in Figure 4 is not shown in Figure 5, and all of the boiling line for pure *b* is hidden in the figure by the two-phase region. Keep in mind the two-phase region continues to lower temperatures, and solids will eventually freeze out as the temperature is lowered.

When the temperature, pressure, and composition are set to a point within the two-phase region, *l-v* phase separation occurs. The compositions of the separate liquid and vapor phases are given by the intersection of a tie line with the boundary surface of the *l-v* region. The tie line runs through the point in question, parallel to the composition axis. The phase that is higher in *a* is the vapor. The compositions of the liquid and vapor are fixed, and the amounts present of each of these phases is determined by the overall composition specified for the system.

Figure 5 happens to be a representation of the CO_2-methanol system. However, the main characteristics of this particular system apply equally well to any Type I mixture, so do not put any generality on the specific temperatures and pressures in this figure. Conventional LC becomes possible for Type I solvent systems when the upper *l-v* phase boundary surface is below atmospheric pressure at ambient temperatures so that the mobile phase is liquid over the entire length of the column (assuming the outlet is at or near atmospheric pressure). For example, methanol-water is a Type I system and can be conveniently used in conventional LC at ambient

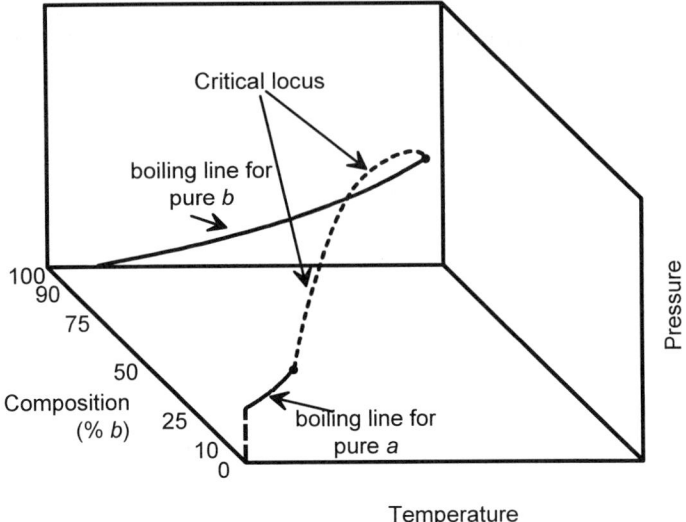

Figure 4. Type I binary mixtures have a locus of critical points spanning the composition dimension and connecting the critical points of the two, pure components in the limiting planes. (Adapted with permission from reference 29.)

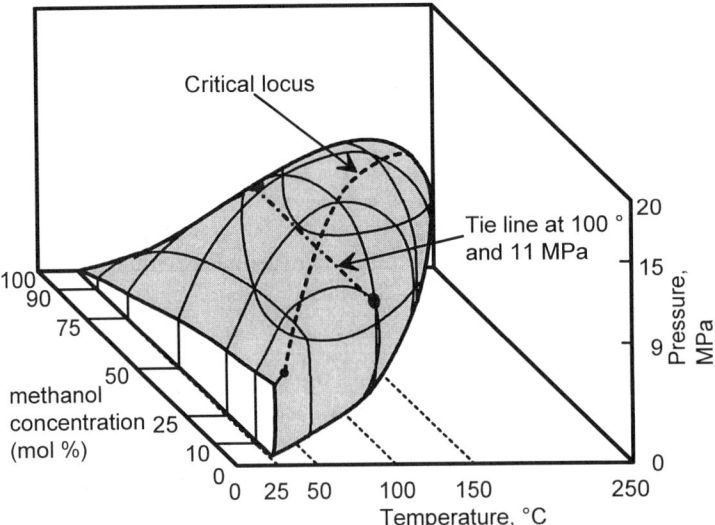

Figure 5. The two-phase *l-v* region of a binary mixture is a volume in a three-dimensional phase diagram. The Type I mixture CO_2-methanol is illustrated here. The two-phase region is the shaded interior of the figure. It has been cut off at 25 °C to show the isotherm, but actually extends to lower temperatures. (Adapted with permission from reference 29.)

temperature and at somewhat elevated temperatures before the mobile phase boils (that is, before the system enters the two-phase l-v region at ambient pressure).

The region outside the shaded section of Figure 5 is a continuous one-phase region analogous to the continuous one-phase region shown in Figure 3 for a pure fluid. It encompasses both the "normal" liquid and vapor states of the binary system. We are free to continuously vary temperature, pressure, and composition within the safety limits of the system, and go anywhere in this fluid continuum without experiencing a phase change as long as we do not enter the two-phase (shaded) region. We also must not forget that solids will freeze out at sufficiently low temperatures. The only restrictions for selecting chromatographic parameters to investigate is that we must stay away from P-T-X coordinates of the phase diagram where the mobile phase is not a single-phase fluid (the shaded section), and we must observe the maximum safe temperature and pressure limits for our system.

Fitting Specific Chromatographic Techniques into the Phase Diagram

We will continue to use the binary phase diagram to describe the l-v behavior of a chromatographic mobile phase without specifically assigning identities to the a and b components. Following the ASTM and IUPAC definitions of a supercritical fluid, we can highlight the section of the binary phase diagram that is always above the critical temperature and critical pressure for the mobile-phase system, that is, the supercritical fluid region for the Type I binary mixture. This is shown in Figure 6. If SFC is the technique practiced when the mobile phase is a supercritical fluid, then this is the region of the phase diagram where SFC is performed. If a one-component mobile phase is used (we will assume it is just the a component of our binary mixture), then the SFC region collapses into the shaded area of the near plane of Figure 6, in agreement with Figure 2. Solute relative retention in SFC depends strongly on the choices for the a and b components, their mole ratio, the temperature, and the pressure (particularly when a is very volatile and the mole fraction of component b is low).

It is important at this point to realize that in packed-column SFC the pressure is usually actively controlled at the column outlet using a backpressure regulator or a pressure-controlled make-up flow (*17*). Mobile-phase flow is controlled at the column inlet. This combination of downstream pressure control and upstream flow control allows volumetric mixing of a and b in a two-pump system, and the simple application of composition gradients in a fashion identical to that in conventional LC. The column inlet pressure cannot be controlled with this arrangement and is always higher than the outlet pressure by the amount necessary to sustain the set mobile-phase flow. In open-tubular SFC, pressure is controlled at the inlet. The flow at the column outlet is mechanically restricted using a porous frit or by tapering the last few millimeters of the column down to a small orifice (*7*). The pressure drop on an open-tubular SFC column is so small (often less than 0.1 MPa) compared to the inlet pressure (often 7-68 MPa) that the pressure is considered controlled at the inlet value over the entire column length. When using packed columns, or any column, under conditions producing a significant pressure drop, it is often the pressure at the column outlet (the point in the column where the mobile phase is weakest and solute retention strongest) that most strongly influences the overall chromatographic behavior.

What technique would we be doing if the temperature remained above the critical locus but the column outlet pressure were below? This region is shown in Figure 7. So far it is unnamed and largely unexplored. Many pressure-programmed SFC procedures are actually started in this region and then elevated into the *formal* SFC region in the course of the separation. If chromatography performed entirely in this lower-than-supercritical-pressure region needed a separate name, we could name it *Hyperbaric Chromatography* (HC). We would predict that solute retention would be much less in this region than when using the same column under GC conditions, but that chromatographic behavior here would be more GC-like than is SFC. Although the solvent strength of any particular mobile phase would be lower in HC than in SFC, the strength would be highly pressure-dependent. Retention and selectivity would also be influenced by the a and b choices, their ratio, and the temperature.

Solvating Gas Chromatography (SGC) can be thought of as an extension of conventional GC in which a mobile phase (usually a single component, CO_2 or NH_3) is chosen that will solvate the sample components to a significant extent at the column inlet pressure. In much of the work reported so far, the outlet pressure of packed microcolumns has been allowed to default to ambient (*18-22*). Thus, SGC could also be considered a limiting case of HC, that is, when the outlet pressure is allowed to default to ambient. The non-uniformities in velocity, mobile-phase strength, and local retention factors in these columns lead to complicated chromatographic behavior that should not be interpreted casually. However, the resolution achievable per unit time has surpassed the capabilities of conventional open-tubular GC in some cases. In our phase-behavior model, SGC would be practiced over the front face of the Hyperbaric Chromatography region with the understanding that the column outlet is at ambient pressure.

You may have figured out that we are headed toward GC behavior, but we will postpone describing GC temporarily and move now to lower temperatures. The naming of techniques performed below the critical temperature has ignored the critical pressure. The realm of Subcritical (or near-critical) Fluid Chromatography (SubFC) is shown in Figure 8, where the temperature is always below the critical locus and the pressures ranges from ambient to the limits of the instrument. The a and b choices and their ratio would have a large influence on solute relative retention and selectivity, as would temperature. However, fluid in this region is less compressible and less pressure-tunable than when in the supercritical fluid region. If advantages arise with the use of low temperatures, a very volatile a choice will allow very low temperatures to be used. CO_2-based mobile phases are routinely used to temperatures as low as -50 °C.

A similar but separately named technique, Enhanced-Fluidity Liquid Chromatography (EFLC), and a subset called Elevated-Temperature EFLC, also exist over essentially the same temperature and pressure range and are shown in Figure 9

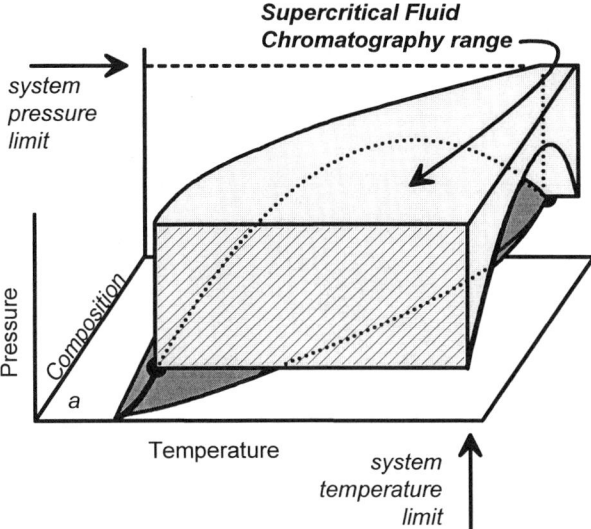

Figure 6. The region of the mobile-phase phase diagram, for Type I binary mixtures, in which supercritical fluid chromatography can be performed according to the formal definition of a supercritical fluid (light shading). The two-phase *l-v* region (dark shading) lies below and left of the SFC region. The near plane in the figure representing 100% component *a* is analogous with Figure 2.

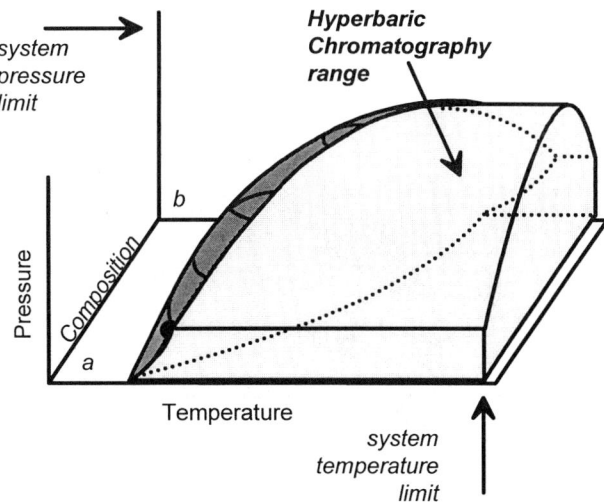

Figure 7. The region for Hyperbaric Chromatography (light shading) where the temperature is above critical and the pressure is below. The two-phase *l-v* region (dark shading) lies to the left.

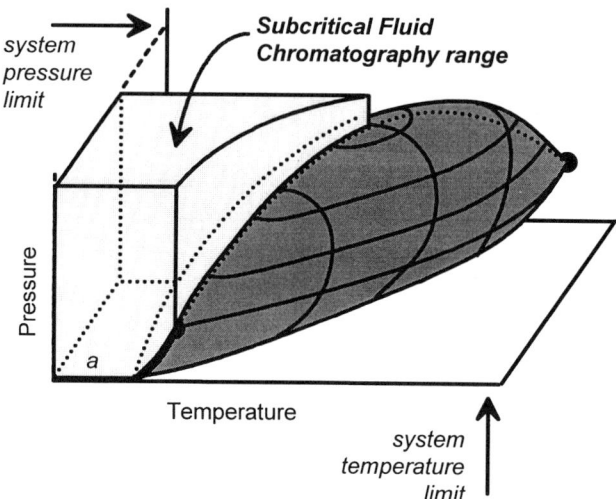

Figure 8. Subcritical (or Near-Critical) Fluid Chromatography (light shading) is performed when the mobile phase is mostly component a at temperatures below critical and at pressures between the vapor pressure and the safety limit. The two-phase l-v region (dark shading) is to the right and below.

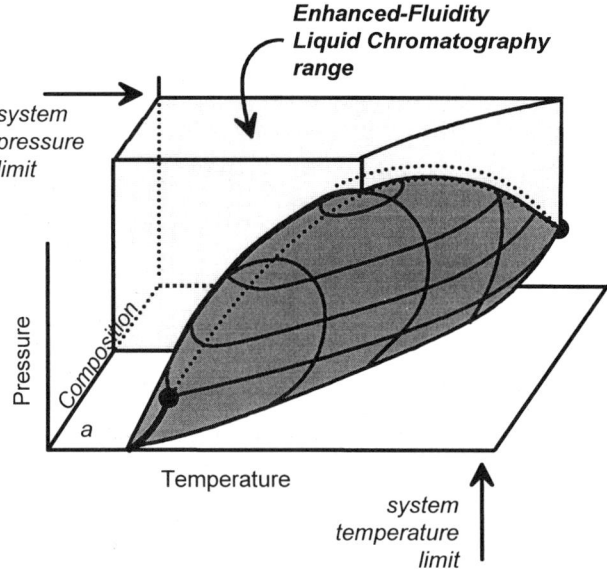

Figure 9. Enhanced-Fluidity Liquid Chromatography (light shading) is much like SubFC except the mobile phase is mostly component b, which is chosen to be an ordinary liquid. The two-phase l-v region (dark shading) is to the right and below.

(23,24). The only difference between EFLC and SubFC is that EFLC is defined as having mostly a "liquid" mobile phase (which would be *b*, our less-volatile component) with a viscosity-reducing component (*a*) added. The purposes of lowering the viscosity include raising diffusion rates, increasing optimum velocities, and shortening analysis times. So, in this phase-diagram model, EFLC occupies the space where the mole fraction of *b* is greater than 0.5, the temperature is subcritical, and the pressure ranges from the mobile-phase vapor pressure to the safety limit for the apparatus. The adjacent realm where the mole fraction of *b* is less than 0.5 is left to be called SubFC. (We should point out that EFLC has frequently been done using ternary mobile-phase systems such as water-methanol-CO_2, where the CO_2 is the viscosity reducer. With sufficient water in such a system, a miscibility gap occurs limiting the amount of CO_2 that can be added without causing a phase separation, thus imposing a real boundary between EFLC and SubFC. However, the phase-diagram model given here is completely accurate for systems that form Type I binary mixtures.) Mobile phases in the EFLC region would behave similarly to those in the SubFC region with respect to the influence of composition, temperature, and pressure, except that we would expect pressure to have an even smaller influence. The compressibility steadily diminishes in this region as we raise the content of a "liquid" *b* component.

In this model, LC and GC are limiting behaviors, and are situated as shown in Figure 10. Conventional LC is usually practiced at or near ambient temperature. Since typical LC liquids have so little compressibility at *conventional* LC pressures, there is no reason to be concerned about pressure except to generate flow (as long as the columns and their packings are not damaged). In spite of the pressure drop over the column length, the strength of the mobile phase is nearly uniform over the entire column in *conventional* LC (in isocratic operation). If flow could be generated by some other means, the relative retention of solutes would be unchanged from the pressure-driven-flow case. Therefore, pressure is not a parameter capable of controlling relative retention to a significant extent in *conventional* LC. There is no particular reason to adjust the outlet pressure, and it is usually allowed to default to ambient. Therefore, in our model, conventional LC exists in the limit when the temperature and outlet pressure are allowed to default to ambient and *a* and *b* are both chosen from among *well-behaved* liquids under these conditions. High-temperature LC basically uses conventional LC as a starting point and moves toward higher temperatures, sometimes with additional outlet pressure applied to somewhat raise the boiling temperature of the mobile phase. (There are *measurable* effects of pressure on mobile-phase strength and selectivity using common liquids at conventional LC pressures (25). Pressures well above those of conventional LC have even more influence (26). But these effects are very small compared to the pressure dependence observed in SFC and EFLC experiments.)

GC behavior exists when solute--mobile-phase forces are essentially zero. Pressure has no influence on relative retention when the intermolecular forces in the mobile phase are zero to begin with. Thus, pressure has no purpose in conventional GC other than to generate flow. The column-outlet pressure is allowed to default to

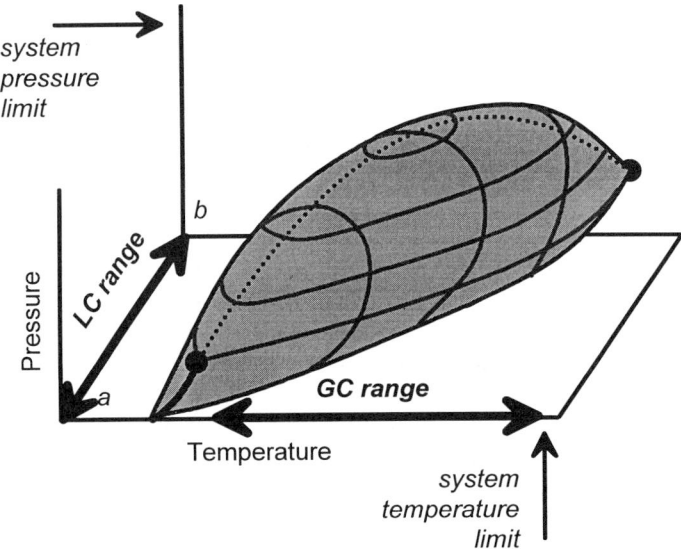

Figure 10. The regions for conventional GC and LC shown relative to the two-phase *l-v* region (dark shading).

whatever is convenient, usually to the atmosphere or to the pressure of a detector. Similarly, the mobile-phase composition has no influence on solute retention if there are no significant intermolecular forces in the mobile phase, so only one mobile-phase component is necessary. It is chosen to be inert to the solutes and to the stationary phase, and to provide fast diffusion rates. Thus, helium and hydrogen are widely used. High-temperature GC simply extends the classical behavior to higher temperatures.

Degrees of Freedom in the Mobile Phase

We have described the mobile phase in terms of its temperature, pressure, and composition. We will now consider the influence of these mobile-phase parameters on solute relative retention and selectivity. Which mobile-phase parameters can we use to adjust solute relative retention and selectivity, and how many ways can we control a separation? In this consideration we will not include stationary phase choice or adjustment, but limit ourselves only to mobile-phase parameters.

In HC at solvating pressures and in SFC at high temperatures and low b concentrations, we would expect strong dependence of solute retention on all three mobile-phase parameters. Thus, we would have three independent means of adjusting retention, i.e., three degrees of freedom, assuming these parameters affect various solutes in different ways. In SFC and in SubFC and EFLC, the influence of pressure diminishes as the concentration of the b component increases and as the temperature decreases. This occurs because the mobile phase becomes less compressible as we go in these directions, but temperature and composition both remain important. We might say that we have something between two and three degrees of freedom under these conditions. (But keep in mind that pressure is still extremely important in keeping the mobile phase from boiling.)

Conventional GC and LC are clearly limited in their behaviors and our ability to make adjustments affecting relative retention. In GC the mobile phase is completely inert and non-solvating, so pressure and composition have no influence on solute relative retention. The only retention-control parameter is temperature; thus GC has one degree of freedom. In conventional LC, mobile phases are nearly incompressible, so pressure has little useful influence on relative retention. Temperature is important, but has been allowed to default to ambient in most of the work reported to date. LC at ambient temperature has just one degree of freedom, the mobile-phase composition. If temperature could be fully exploited we might say that LC has two degrees of freedom, but the useful temperature range is very limited as long as the outlet pressure is allowed to default to ambient. Clearly, we have to control the column-outlet pressure in LC to achieve the most benefit from temperature tuning.

Unification

You may have already noticed that the various individually named chromatography regions within the phase diagram model fit together perfectly, with no gaps and with no meaningful physical or chemical discontinuities, filling the continuum of one-phase fluid behavior described earlier. This is depicted in Figure 11. The only things separating these techniques are their arbitrary names. Clearly, the boundaries separating these individually named techniques are completely without merit except for the surface of the 2-phase l-v region. This is unified chromatography from the mobile phase perspective. *We are completely free to use conditions anywhere within the single-phase continuum within the pressure and temperature limits of our instrument.* Furthermore, we can take full advantage of favorable mobile-phase properties that do not exist in the limiting, conventional techniques of GC and LC.

Building a Unified Chromatograph. Let us omit conventional GC from the picture for the moment (because of the practical difficulties of designing one pump that could operate equally well with helium or hydrogen, with liquefied gases, and with normal liquids). We will now consider the essential capabilities necessary for a single, packed-column instrument to use the remainder of the continuum. We need a pumping system with two independent pumps operated under flow control to allow volumetric mixing of the a and b solvents. The a pump would have to be capable of pumping fluids as volatile as liquid CO_2 or as ordinary as any normal liquid ranging from water to hexane. The b pump could be similar or could be a conventional LC pump if we are willing to restrict the b solvent to ordinary liquids.

A high-pressure loop injector is completely sufficient. A temperature controller is necessary for the column. It is not particularly important for the injector to be at the column temperature as long as the sample is soluble in the mobile phase at ambient temperature and there is continuity of the fluid phase behavior between the injector and the column inlet (going through the temperature change). The column could be a perfectly ordinary LC-type column, but we must remember the temperature cannot exceed the maximum dictated by the column components and the stationary phase. Selecting stable columns and stationary phases will increase our usable temperature range.

For now we will consider using a flow-through spectroscopic detector. It too can be a perfectly ordinary LC detector in its spectroscopic traits. Just as with the injector, the detector can operate at ambient temperature as long as the sample components remain soluble in the mobile phase and there is continuity of the fluid phase between the column outlet and the detector. We can't ignore pressure in the detector, however, since our mobile phase may be highly volatile and may boil at detector temperature (even if it is only ambient) if we don't keep the pressure sufficiently elevated. For this reason we will operate the detector at column outlet pressure rather than at ambient pressure. We can accomplish this by controlling the pressure downstream from the detector instead of at the column outlet. For this we need to add a backpressure-regulating device as described earlier.

We would want wide specifications for this instrument to allow us a generous coverage of the mobile-phase continuum. Continuously adjustable column temperature from -60 °C to 350 °C, and outlet pressure from ambient to about 40-45 MPa are easily achievable with today's technology.

This instrument we just described, a *Unified Chromatograph*, is shown schematically in Figure 12. This instrument would be capable of operating over the entire mobile-phase continuum (except for conventional GC). This encompasses conventional LC, SubFC, EFLC, SFC, and HC. It only excludes conventional GC because of the pump restriction we imposed, not because of any discontinuity in fluid behavior.

This instrument exists today. It is available commercially in analytical scale from three different suppliers (although not all three provide quite the full temperature and pressure range mentioned earlier), and is called a *Packed-Column Supercritical Fluid Chromatograph*. This is a most unfortunate name choice since it implies limited capability (as described by Figure 5) while the actual instrument is considerably more capable than a conventional LC instrument! In fact, looking at Figure 12 we see that all the components of a conventional LC instrument are included. This, of course, is a consequence of our model since LC is a limiting case of Unified Chromatography, and LC is not excluded by our pump specifications.

We could go through a similar exercise to design a unified chromatograph for open-tubular chromatography. In this case we might include GC (because of the strong similarity of existing open-tubular GC and SFC instruments) but omit LC. This choice would arise because open-tubular LC is too slow to interest practical analysts.

Advantages of Unified Chromatography

Let us assume we are separating two dissimilar solutes that co-elute under a particular set of mobile-phase conditions. The specific intermolecular forces responsible for retaining these solutes are likely going to be different for each solute (even though the total influence of all these forces is the same since they coelute). The two solutes are therefore likely to respond somewhat differently to changes in the *a/b* ratio, the temperature, and the pressure if we work in a region where these parameters have influence, as described earlier. Thus, we have opportunities to adjust the selectivity and to accomplish a reasonable separation by varying easily and rapidly adjustable mobile-phase parameters. Figure 13 shows an SFC example of the influence of temperature on a separation of vitamins and how selectivity can be effectively tuned *(27)*. This mobile-phase tunability decreases methods development time because temperature and pressure can be changed and the column re-equilibrated much quicker than columns and mobile-phase components can be changed. Fewer columns and mobile-phase components need to be investigated before arriving at a suitable separation.

Reducing mobile-phase viscosity and increasing mobile-phase diffusion rates result in faster optimum velocities and shorter analyses than in conventional LC. Picking a highly volatile *a* component, minimizing the %*b*, and using relatively high

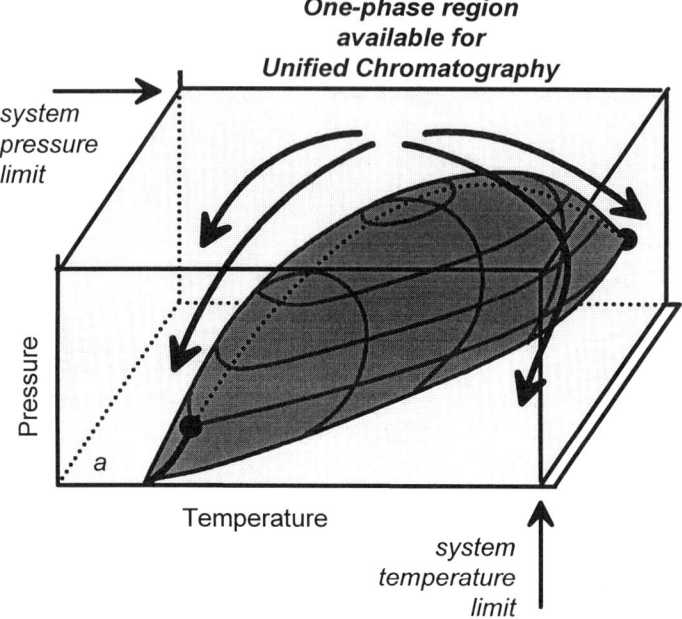

Figure 11. Unified Chromatography is the combination of all the individually named techniques over the continuum of single-phase fluid behavior. We are free to set pressure, temperature, and composition anywhere in the light-shaded region without regard to technique naming conventions. We must avoid the darkly shaded region where l-v separation occurs.

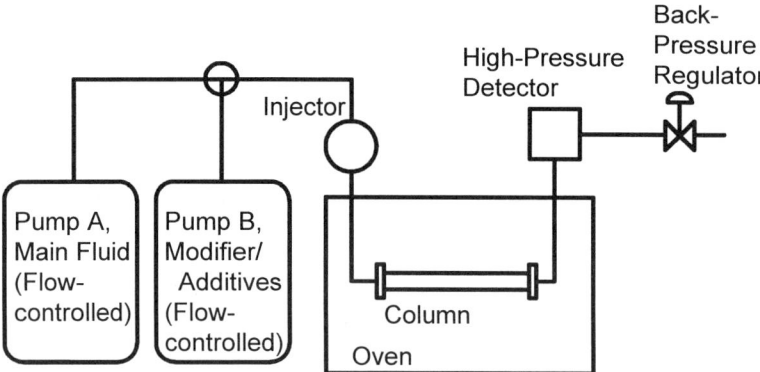

Figure 12. A packed-column Unified chromatograph. This type of instrument is already available commercially and is known as a packed-column Supercritical Fluid Chromatograph, an unnecessarily limiting misnomer.

temperatures and low pressures all increase speed. Optimum velocities up to about ten-times conventional LC are possible, but factors in the range of three to five are quite realistic and can usually be achieved.

With much faster optimum velocities, analysts can often afford to use higher retention factors while still accomplishing a separation in less time than in conventional LC. This can significantly contribute to resolution in cases where selectivity between solutes is low (that is when the separation factor, α, is near unity). Another possibility afforded by lowering the viscosity is the ability to use much longer columns than is possible in conventional LC. Again, this ability helps us in situations where the solutes have low α values, are retained well, and the only recourse in achieving additional resolution is to add theoretical plates (column length) to the system. While lengthening the column obviously adds more time to the analysis, the analysis time with 1-m-long columns in packed-column SFC is approximately the same as with a 0.25-m column used with conventional LC. A separation helped by using a long column is shown in Figure 14 (*28*).

In more ordinary separations, the operating costs per sample can be reduced tremendously by minimizing analysis time and achieving more work per analyst, instrument, square meter of lab space, and etc. The use of CO_2 as a mobile-phase component is not required in the general concept of Unified Chromatography, but when it can be used additional savings are possible by reducing the use of liquid solvents and the production of liquid waste.

Optimization in Unified Chromatography

This is conceptually quite simple. After picking the stationary phase and the *a* and *b* solvents to investigate, a worker simply varies the remaining parameters in a systematic fashion to find the global optimum within our permissible parameter range, that is the mobile-phase continuum limited by the system maximum temperature and pressure. Note that this is no more complicated than adding column temperature and outlet pressure to what is already done to optimize LC separations. The starting point can be anywhere in the fluid continuum as long as the optimization process is wide enough to find the global optimum. The driver for the optimization would be specified according to the worker's immediate needs, such as achieving at least some required resolution between sample components in minimum time.

Finding the global optimum assures us that we have found the best separation possible for the initial stationary phase and solvent choices according to the needs specified by the driver. Note that professional analysts would not care in the least where the optimum exists in the phase diagram or what the optimum technique would be called according to anyone's naming convention. The technique name simply does not matter in the unified approach. What is important to the analyst is searching the entire parameter space and finding the *best* separation conditions rather than settling for any ambient condition by default.

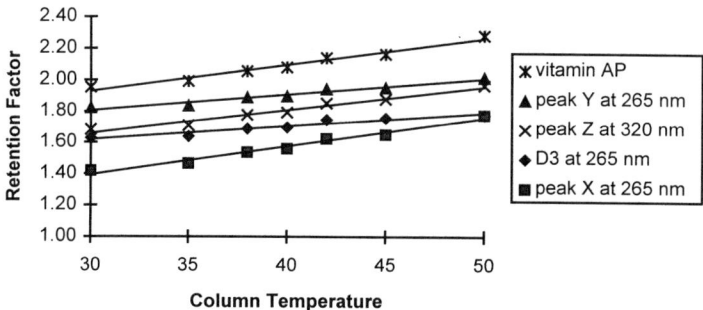

Figure 13. Selectivity tuning by temperature for vitamins A palmitate and D3. 5-μm Inertsil Phenyl, 4.6 mm x 250 mm; 2% methanol in CO_2 run at 18 MPa outlet pressure. Peak D3 co-elutes with peak z at 30 °C, and with peak x at 50 °C, but is baseline-resolved from both at 40 °C (27).

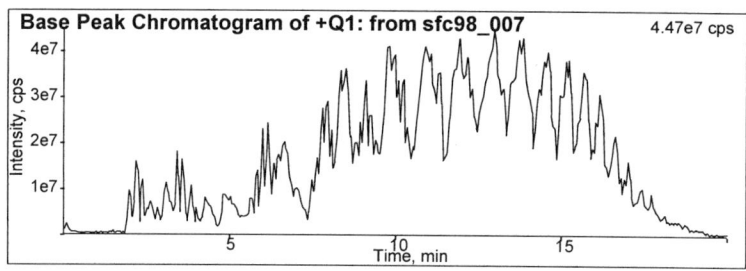

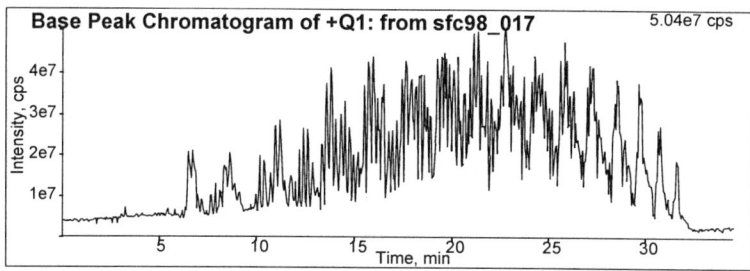

Figure 14. Comparing a 250-mm-long column (upper) with a 750-mm-long column (lower), both used at 2 mL/min, for the separation of a complicated nonionic surfactant mixture. Conditions: 4.6-mm x 250-mm Deltabond cyano column (three used in series in the lower figure), 100 °C, 20 MPa outlet pressure. CO_2/methanol gradients: 3-minute initial hold at 1 (volume) % methanol followed by a 1%/min ramp to 10% methanol (upper), and a 3-min initial hold at 3% methanol followed by a 0.5 %/min ramp to 30% methanol (lower) (28).

It is important to be completely fair and to point out that if the a and b solvents are both normal liquids, outlet pressure may not be especially important if the optimum temperature is well below the normal boiling point of the mobile phase. In these cases ambient outlet pressure is as good as any and is extremely convenient. However, with outlet pressure control it will likely be rare that normal liquids will provide the performance (particularly speed of analysis) capable when at least one mobile-phase component is too volatile for conventional LC.

Cost and Risk

The additional components necessary for a Unified Chromatograph are relatively inexpensive and only add about 20% to the price of an otherwise comparable LC instrument. Some improvements in the a pump are necessary to allow pumping very volatile fluids and liquefied gases. Improvements are also required in the column oven and the detector cell (to accommodate detection at the column-outlet pressure). Finally, it is necessary to add a downstream pressure regulator. These additional controls add tremendous capability above and beyond basic LC and compromise absolutely nothing.

With this in mind, consider the risk vs. the reward of Unified Chromatography. If a worker is already considering the purchase of a conventional LC instrument, the only money put at risk is the 20% addition for the new capabilities, not the entire cost of the instrument. This is so because, in the worst possible outcome, the worker would still have all the capabilities (and more) of a brand new LC instrument within the Unified Chromatograph. Considering rewards, the worker may increase the rate of producing results by three to five times and reduce the cost per analysis by as much as 70%. In addition, the fuller range of selectivity and its control and the additional efficiency via longer columns, all possible in Unified Chromatography, will lead to successful separations not always possible with conventional LC.

Conclusion

With the unified picture in mind, it becomes apparent that arguments aimed at discrediting any specifically named technique while benefiting another are totally baseless. For example, it makes no sense to condemn SFC while defending conventional LC or GC. Likewise, it makes no sense to promote the somewhat limited expansion of conventional techniques, for example high-temperature LC with ambient or slightly elevated outlet pressure, while discrediting new techniques with really expanded capabilities.

The defense of established, named techniques is a natural reflex resulting from several causes. Perhaps the most significant is the research funding mechanisms existing in most of the world. Established researchers often must defend their vested interests or risk losing their funding. New researchers have to find a unique, new technique to define, promote, and defend in order to get and keep funding. Thus, amplifying trivial differences and building up otherwise meaningless walls between similar techniques has occurred for many years. This will be really difficult to undo.

Instrument manufacturers have also adopted strategies reinforcing the proliferation of individually named but similar techniques. The manufacturers' first interest is to defend their existing instrument market. When a new technique first appears, it is usually condemned by all those manufacturers feeling threatened. Those manufacturers offering something radically new often position it not to threaten the other products they offer. Thus, new named techniques are created, and special, new instruments or accessories are sold. The manufacturers following this divide-the-market strategy would like nothing better than to sell every well-equipped laboratory at least one of every kind of instrument they can name. This is the natural course until the customers catch on, change their expectations, and the manufacturers respond to the demands of the market.

Unified Chromatography, as described here, puts an end to all this. It greatly widens the possibilities for choosing mobile-phase components, temperatures, and pressures; focuses on achieving the best results possible wherever they might occur within the mobile-phase continuum; and represents no loss or compromise in any existing capability.

Literature Cited

1. Antia, F. D.; Horvath, Cs. *J. Chromatogr.* **1988**, *435*, 1-15.
2. Snyder, L. R.; Dolan, J. W.; Molnar, I.; Djordjevic, N. M. *LC-GC* **1997**, *15*, 136-151.
3. Carr, P. W.; Mao, Y.;McNeff, C. V.; Li, J. *Advantages of Ultra Stable Stationary Phases for RPLC at Temperatures Approaching 200 °C*, Presented at the Thirty-Seventh Annual Eastern Analytical Symposium and Exposition, Somerset, NJ, U.S.A., November 15-20, 1998, Paper 177.
4. Liu, G.; Djordjevic, N. M.; Erni, F. *J. Chromatogr.* **1992**, *592*, 239-247.
5. Trones, R.; Iveland, A.; Greibrokk, T. *J. Microcolumn Sep.* **1995**, *7*, 505-512.
6. *Analytical Supercritical Fluid Chromatography and Extraction*; Lee, M. L., Markides, K. E., Eds.; Chromatography Conferences: Provo, UT, 1990.
7. Chester, T. L. Supercritical Fluid Chromatography Instrumentation. In *Analytical Instrumentation Handbook*, 2nd Edition; Ewing, G. W., Ed.; Marcel Dekker: New York, NY, 1997; 1287-1350.
8. *Supercritical Fluid Chromatography*, Smith, R., Ed.; RSC Chromatography Monographs; The Royal Society of Chemistry: London, UK, 1988.
9. *Fractionation by Packed-Column SFC and SFE—Principles and Applications*, Saito, M., Yamauchi, Y., Okuyama, T., Eds.; VCH Publishers, Inc., New York, NY, 1994.
10. Berger, T. A.; Packed Column SFC, RSC Chromatography Monographs; The Royal Society of Chemistry: Cambridge, UK, 1995.
11. Supercritical Fluid Chromatography with Packed Columns—Techniques and Applications, Anton, K., Berger, C., Eds.; Chromatographic Science Series; Marcel Dekker, Inc., New York, NY, 1998.
12. Chester, T. L. *Anal. Chem.* **1997**, *69*, 165A-169A.

13. National Fire Protection Association. http://www.nfpa.org (accessed Feb 1998).
14. "Standard Guide for Supercritical Fluid Chromatography Terms and Relationships"; Designation E 1449-92; Annual Book of ASTM Standards; American Society for Testing and Materials, Philadelphia, PA, 1995; Vol. 14.02, 905-910.
15. Smith, R. M., *Pure & Appl. Chem.* **1993**, *65*, 2397-2403.
16. van Konynenburg, P. H.; Scott, R. L. *Philos. Trans. R. Soc. London, Ser. A.* **1980**, *298*, 495-540.
17. Chester, T. L.; Pinkston, J. D. *J. Chromatogr. A*, **1998**, *807*, 265-273.
18. Shen, Y.; Lee, M. L. *J. Chromatogr. A*, **1997**, *778*, 31-42.
19. Shen, Y.; Lee, M. L. *Anal. Chem.* **1997**, *69*, 2541-2549.
20. Shen, Y.; Lee, M. L. *Chromatographia* **1997**, *46*, 537-544.
21. Shen, Y.; Lee, M. L. *Chromatographia* **1997**, *46*, 587-592.
22. Shen, Y.; Lee, M. L. *Anal. Chem.* **1998**, *70*, 737-742.
23. Cui, Y.; Olesik, S. V. *Anal. Chem.* **1991**, *63*, 1812-1819
24. Lee, S. T.; Olesik, S. V. *Anal. Chem.* **1994**, *66*, 4498-4506.
25. McGuffin, V. L., Evans, C. E., and Chen, S. *J. Microcolumn Sep.* **1993**, *5*, 3-10.
26. MacNair, J. E., Lewis, K. C., Jorgenson, J. W. *Anal. Chem.* **1997**, *69*, 983-989.
27. Hentschel, R. T.; Chester, T. L. *Final Program and Book of Abstracts*, 8th International Symposium on Supercritical Fluid Chromatography and Extraction, St. Louis, MO, U.S.A., July 12-16, 1998; p E-17.
28. Pinkston, J. D.; Baker, T. R. *Final Program and Book of Abstracts*, 8th International Symposium on Supercritical Fluid Chromatography and Extraction, St. Louis, MO, U.S.A., July 12-16, 1998; p E-20.
29. Chester, T. L. *Microchem. J.* **1999**, *61*, 12-24.

Chapter 3

Pressure as a Unifying Parameter in Chromatographic Retention

C. E. Evans, M. C. Ringo, and L. M. Ponton

Department of Chemistry, University of Michigan, Ann Arbor, MI 48109-1055 (ceevans@umich.edu)

As a fundamental thermodynamic parameter, pressure may be expected to play an important role in solute retention for all chromatographic separations. However, pressure is commonly considered to be a significant parameter in chromatographic separations only when the mobile phase is highly compressible. In this chapter, the impact of pressure on solute retention is assessed for gas, supercritical fluid, and liquid chromatography. Contributions from both the bulk-phase compressibility and changes in partial molar volume upon interaction are evaluated. Although the pressure dependence of solute retention in LC is commonly overlooked due to minimal mobile-phase compressibility, changes in the solute partial molar volume upon interaction with the mobile and stationary phases give rise to significant pressure-induced capacity factor shifts. Together, these factors point out the important unifying role of pressure as a fundamental parameter in describing chromatographic retention.

Introduction

Solute retention in chromatographic separations results from the normalized migration rate of solutes through a column containing a mobile and a stationary phase. As a result, solute retention is a complex function of solute-solvent interactions in the mobile phase and solute-interphase interactions in the stationary phase. Several generalized, unified models have been proposed to describe solute retention (1-5). As a fundamental thermodynamic parameter, pressure is expected to be directly related to solute retention for all chromatographic separations. However, pressure is often considered to be an important parameter only for those separations where the mobile-phase compressibility is considerable. Indeed, the relationship between pressure and solute retention has been studied extensively for gas and supercritical fluid separations. Only recently has pressure been shown to be an important parameter in liquid chromatographic separations. In this manuscript, we illustrate pressure as a unifying parameter in chromatographic separations - showing the importance of bulk-phase compressibility as well as solute partial molar volumes.

Theoretical Considerations

Although chromatographic separation is an inherently nonequilibrium process, solute retention is evaluated at the centroid of the solute zone where near-equilibrium conditions are maintained. As a result, solute retention may be characterized using equilibrium thermodynamics (5). One primary descriptor of solute retention in elution chromatography is capacity factor (k), which relates the retention time of a solute to the void time of a nonretained compound. Solute capacity factor is directly related to the equilibrium distribution coefficient of the solute between the stationary and mobile phases (K), and the volume ratio of the stationary (V_{stat}) and mobile (V_{mob}) phases.

$$k = K \frac{V_{stat}}{V_{mob}} \qquad (1)$$

As a primary parameter, pressure may be expected to have a significant impact on both the bulk properties and the equilibrium properties of chemical systems (6, 7). As a result, the equilibrium constant as well as the absolute volumes dictating the phase ratio may be pressure dependent. However, as chromatographic systems are commonly fixed volume, an absolute volume decrease in the more compressible mobile phase must be mirrored by an absolute volume increase in the stationary phase. As a result, the overall, pressure-induced change in the absolute volumes of the phases will be limited by the least compressible phase. In most cases of chromatographic interest, the pressure dependence of the phase ratio will be limited by the stationary phase compressibility. Although the stationary phase has been shown to exhibit measurable changes in volume in some cases (*vide infra*), the resulting contribution to the pressure-induced perturbation in solute retention is quite modest. In contrast with absolute phase volumes, a wide range of equilibria have been shown to be pressure dependent (6, 7). As a result, the pressure dependence of solute retention is most often predominated by pressure-induced shifts in separation equilibria.

Assuming an immiscible stationary and mobile phase, previous thermodynamic predictions have shown that solute retention can be described as a function of a partial molar volume term and a bulk compressibility term (8-11).

$$\left(\frac{\partial \ln k}{\partial P}\right)_T = -\left(\frac{\overline{V}_{i,stat} - \overline{V}_{i,mob}}{RT}\right) - \kappa_{mob} = -\frac{\Delta \overline{V}_i}{RT} - \kappa_{mob} \qquad (2)$$

The validity of this expression depends on the presumption of no solute-solute interactions and describes a fixed volume system where $V_{mob} + V_{stat}$ = constant. The first term describes the difference in the solute partial molar volume between the mobile phase ($\overline{V}_{i,mob}$) and the stationary phase ($\overline{V}_{i,stat}$). The solute partial molar volume ($\overline{V}_i$) is given by

$$\overline{V}_i = \left(\frac{\partial V}{\partial n_i}\right)_{T,P} \qquad (3)$$

where V is the solution volume and n_i is the number of moles of solute. Solute-solvent or solute-interphase interactions have a profound influence on the volume per mole of solute. That is, cohesive solute-solvent interactions lead to negative values for $\overline{V}_{i,mob}$, whereas repulsive interactions result in positive $\overline{V}_{i,mob}$ values. The second term in eq 2 describes the relative decrease in mobile-phase volume with pressure as the isothermal compressibility, κ_{mob}.

$$\kappa_{mob} = -\frac{1}{V}\left(\frac{\partial V}{\partial P}\right)_T \tag{4}$$

As shown in Table I, the fluids of interest here exhibit a broad range of isothermal compressibilities, spanning about 4 orders of magnitude. Based on eq 2, the expected impact of mobile-phase compressibility and solute partial molar volume differences on solute capacity factor is illustrated in Table II. Clearly, for those cases where both the mobile-phase compressibility and the $\Delta \overline{V}_i$ are considerable, there is a large impact on solute capacity factor. Interestingly, even for cases where the mobile-phase compressibility is modest, significant perturbations in solute capacity factor are predicted for some solutes.

Table I. Relative contributions of mobile-phase compressibility and solute partial molar volumes to the pressure dependence of solute retention.

	κ_{mob}	$\left(\dfrac{\overline{V}_{i,stat} - \overline{V}_{i,mob}}{RT}\right)$	$\left(\dfrac{\partial \ln k}{\partial P}\right)_T$
GC	10^{-1} bar^{-1} significant contribution	$\overline{V}_{i,mob} \sim 0$; $\overline{V}_{i,stat}$ moderate small contribution	Magnitude: moderate Direction: negative
SFC	10^{-1} to 10^{-3} bar^{-1} significant contribution	$\overline{V}_{i,mob}$ dominates very large contribution	Magnitude: very high Direction: negative
LC	10^{-4} to 10^{-5} bar^{-1} negligible contribution	$\overline{V}_{i,mob}$ and $\overline{V}_{i,stat}$ small significant contribution	Magnitude: moderate Direction: negative or positive

Table II. Predicted pressure-induced changes in solute capacity factor for a 100 bar increase in pressure.[†]

κ_{mob}	$\Delta \overline{V}_i$	$\Delta k/k$ (%)
10^{-4} bar^{-1}	---	-1%
10^{-2} bar^{-1}	---	-60%
10^{-4} bar^{-1}	+10 cm^3/mol	-5%
10^{-4} bar^{-1}	+100 cm^3/mol	-30%
10^{-4} bar^{-1}	-10 cm^3/mol	+3%
10^{-4} bar^{-1}	-100 cm^3/mol	+50%
10^{-2} bar^{-1}	+100 cm^3/mol	-80%
10^{-2} bar^{-1}	+1000 cm^3/mol	-100%

[†] $T = 298$ K.

Originally derived for supercritical fluid applications, the relationship shown in eq 2 does not require specification of the mobile phase and is applicable to gas, liquid, or supercritical fluid phases. In this chapter, we compare and contrast the pressure dependence of solute retention for the more well-known gas and supercritical fluid mobile phases with the less characterized liquid mobile phases. Specific attention is focused on the relative roles of mobile-phase compressibility and changes in solute partial molar volume as shown in Table I.

Gas and Supercritical Fluid Chromatography

The pressure dependence of solute retention has been extensively studied for both gas (*12-15*) and supercritical fluid (*16-27*) chromatography. Considering first the influence of bulk-phase behavior, the isothermal compressibility is quite high for both gases and supercritical fluids (Table I). As a result, the mobile-phase compressibility represents a significant contribution to the pressure dependence of solute retention in both GC and SFC. Not only is the magnitude of κ_{mob} quite high, but both fluids exhibit a significant range of compressibilities with pressure. Based on eq 2, a pressure increase is predicted to result in a decrease in solute retention if mobile-phase compressibility is the primary contribution.

However, the role of the solute partial molar volumes in the mobile and stationary phases must also be considered. In the case of gases, relatively little interaction between the solute and the gaseous mobile phase is expected, resulting in a $\overline{V}_{i,mob}$ of near zero. Stationary-phase interactions are expected, however, leading to relatively small values of $\overline{V}_{i,stat}$ (*11*). This combination of factors results in a moderate pressure dependence on solute retention in gas chromatographic separations, where the mobile-phase compressibility is the primary contribution. As a result, an increase in pressure gives rise to a modest decrease in solute retention in gas chromatographic separations.

The high compressibility of supercritical fluid phases suggests a dominant role for κ_{mob} in the pressure dependence of solute retention in SFC as well. However, in contrast with gases, supercritical fluids exhibit large changes in solute partial molar volume upon partitioning into the stationary phase. Indeed, the magnitude of the partial molar volume for solutes in the SFC stationary phases is significant. For example, in the polysiloxane stationary phases extensively studied for SFC, naphthalene is shown to exhibit partial molar volumes ranging from +400 to -2000 cm^3/mol (*10, 11, 27*). These values are indicative of the order of magnitude of $\overline{V}_{i,stat}$ with the range likely reflecting differences in the polarity, film thickness, and crosslinking of the stationary phase. By contrast, the partial molar volume of solutes in supercritical mobile phases are nearly always negative and very large in magnitude. Supercritical fluids form a highly compact solvation shell around the solute, resulting in this significant decrease in volume when the solute enters the supercritical phase. Accordingly, partial molar volumes of solutes in supercritical fluids have been determined up to -18,000 cm^3/mol for typical SFC temperatures and pressures, with $\overline{V}_{i,mob}$ approaching extremely large negative values near the critical point (*8, 10, 11, 18, 22, 27*). As a result, pressure-dependent retention in SFC separations is most often dominated by the partial molar volume term in eq 2. In addition, the compressibility and the partial molar volume terms have identical directionality, leading to a large decrease in the solute capacity factor with pressure. It is this large, solute-specific shift in retention that leads to the successful implementation of pressure as a control parameter in supercritical separations.

In contrast to other chromatographic methods, the stationary and mobile phase in supercritical fluid chromatography often exhibit significant mutual solvation. Stationary-phase "swelling" by supercritical fluids can be quite significant, with changes in volume of up to 190% (*19, 21*). These changes in the stationary-phase volume may be pressure dependent, leading to an additional factor affecting the pressure dependence of solute retention. For example, (poly)dimethylsiloxane rubber in contact with supercritical carbon dioxide at 35 °C exhibits a relative change in volume of 50% at 80 bar and 83% at 270 bar (*21*). As a result, supercritical fluid chromatography represents an interesting case in which the stationary-phase volume is not only pressure-dependent, but may increase with increasing pressure. A number of theoretical models for SFC retention have incorporated the influence of stationary phase swelling on the pressure dependence of solute retention, (*1, 2, 9*) with the thermodynamic approach described in eq 2 now including an additional additive term (*9*). Although important, this factor represents only a modest contribution to the overall change in solute retention, contributing less than 10% of the total change in capacity factor with pressure predicted for supercritical CO_2 separations using a poly(dimethylsiloxane) stationary phase(*9*). Thus, the mobile-phase compressibility and differences in the solute partial molar volume are the primary contributors to pressure-induced retention shifts in SFC.

Liquid Chromatography

In contrast with GC and SFC, pressure is commonly ignored in liquid chromatographic separations and considered only as a mechanism to drive mobile-phase flow. Indeed, the isothermal compressibility is several orders of magnitude lower for liquids (Table I) and κ_{mob} does not change significantly over the pressure range commonly utilized in LC (> 350 bar) (*28*). As a result, a 100 bar increase in pressure is predicted to only yield a decrease in k of 1%, based solely on mobile-phase compressibility ($\kappa_{mob} \sim 10^{-4}$ bar^{-1}). Nonetheless, experimental studies have demonstrated that pressure can have a significant impact on solute retention (*29-40*).

Although compressibility is not a major contributor to pressure-induced changes in k, partial molar volume effects have been shown to play an important role. Compared to other separation methods, liquid chromatography employs a diverse number of stationary and mobile phases. Accordingly, a number of different equilibria play a role in solute retention and must be considered in the effect of pressure on retention. Pressure has been implicated in solute ionization changes (*33*) as well as stationary-phase modification by the mobile phase (*31,32*). In one of the earliest studies, ion-exchange interactions are shown to exhibit a positive $\Delta \overline{V}_i$ of 5 cm^3/mol, measured in the region of 500 to 1500 bar for azo dyes separated on a silica stationary phase (*29, 30*). In contrast, HPLC separations in which complexation with a surface-bound species is the dominant retention mechanism, changes in partial molar volume upon complexation are highly solute dependent and may be positive, negative, or zero. For β-cyclodextrin stationary phases, changes in the partial molar volume for solute complexation range from -12 to +17 cm^3/mol for different solutes (*37, 38*). This range from positive to negative values has been hypothesized to arise from differences in the complex structure (*38*). Consistent with their use in liquid chromatography, one of the most investigated classes of retention equilibria is partitioning into an alkylsilane stationary phase. For a series of fatty acid derivatives separated on a monomeric octadecylsilane stationary phase, the change in partial molar volume ($\Delta \overline{V}_i$) upon partitioning is determined to range from -8 to -20 cm^3/mol (*34, 35*). In this case, the alkyl chain of the fatty acid partitions from a pure methanol mobile phase into a C18 stationary phase yielding a $\Delta \overline{V}_i/CH_2$ of approximately -1.2 cm^3/mol (*34, 36*). Recent studies indicate that the isomers of nitrophenol exhibit $\Delta \overline{V}_i$ values ranging from -6 to -8 cm^3/mol, similar in magnitude to their six-carbon alkyl chain counterparts (*39*). These values are also in reasonable agreement with recent studies of dihydroxybenzenes using a nonporous stationary phase and a much broader pressure range (4 kbar) (*40*). Calculations based on retention data for catechol, resorcinol, and hydroquinone, separated using a 10:90 acetonitrile:water mixture, show $\Delta \overline{V}_i$ values of -6, -4, and -10 cm^3/mol, respectively. Interestingly, polymeric stationary phases appear to exhibit a more negative $\Delta \overline{V}_i$ for partitioning, ranging from -27 to -93 cm^3/mol for the derivatized fatty acids (*35*). This observation is consistent with supercritical studies which indicate highly negative values for $\overline{V}_{i,stat}$ for the polymeric phases commonly used in SFC.

In contrast with GC and SFC, mobile-phase compressibility is of little importance in the pressure dependence of solute retention in LC. As a result, the pressure dependence arises solely from differences in the partial molar volume of the solute between the mobile and stationary phases. Thus, although not of the magnitude for supercritical separations, the pressure dependence of retention for LC incorporates the highly solute specific nature found in SFC. This important character may lead to $\Delta \overline{V}_i$ values ranging from positive to negative within the same separation. As a result, pressure-induced changes in solute selectivity may be expected. Thus, although mobile-phase compressibility is minimal, differences in the solute partial molar volume can lead to significant pressure-induced shifts in solute retention in LC.

Implications

An important thermodynamic parameter, pressure is often overlooked in those cases where the mobile-phase compressibility is modest. However, changes in the solute partial molar volume may give rise to significant pressure-induced perturbations in solute retention. This general approach is applicable for all fluid conditions, regardless of the bulk-phase compressibility. In this way, pressure provides a unifying parameter for understanding solute retention in all chromatographic separations. Moreover, these studies have important pragmatic implications. Not commonly considered important in liquid chromatography, pressure is not presently monitored or recorded on a routine basis. However, based on the studies shown here, it would be prudent to incorporate pressure into routine quality control measurements. This practice would be especially valuable for particularly difficult or critical separations. Finally, the influence of pressure on the other chromatographic figures of merit should also be considered. Addressed here only for solute retention, pressure may be an important unifying parameter in assessing solute selectivity and efficiency as well.

References

1. Martire, D. E.; Boehm. R. E. *J. Phys. Chem.* **1987**, *91*, 2433.
2. Martire, D. E. *J. Liq. Chromatogr.* **1988**, *11*, 1779.
3. Dill, K. A. *J. Phys. Chem.* **1987**, *91*, 1980.
4. Dorsey, J. G.; Dill, K. A. *Chem. Rev.* **1989**, *89*, 331.
5. Giddings, J. C. *Unified Separation Science*; Wiley: New York, 1991.
6. Hamann, S. D. *Physico-Chemical Effects of Pressure*; Butterworths: London, 1957.
7. Hamann, S. D. In *Modern Aspects of Electrochemistry*; Conway, B. E. and Bockris, J. O. M., Eds.; Plenum: New York, 1974, Vol. 9; pp. 47-158.
8. van Wasen, U.; Swaid, I.; Schneider, G. M. *Angew. Chem. Int. Ed. Engl.* **1980**, *19*, 575.
9. Roth, M. *J. Phys. Chem.* **1990**, *94*, 4309.
10. Yonker, C. R.; Smith, R. D. *J. Phys. Chem.* **1987**, *91*, 3333.
11. Yonker, C. R.; Smith, R. D. *J. Chromatogr.* **1988**, *459*, 183.
12. Giddings, J. C. *Sep. Sci.* **1966**, *1*, 73.
13. Giddings, J. C.; Myers, M. N.; McLaren, L.; Keller, R. A. *Science* **1968**, *162*, 67.
14. Wicar, S.; Novák, J.; Drozd, J.; Janák, J. *J. Chromatogr.* **1977**, *142*, 167.

15. Pazdernik, O.; Schneider, P. *J. Chromatogr.* **1981**, *207*, 181.
16. Peaden, P. A.; Lee, M. L. *J. Chromatogr.* **1983**, *259*, 1.
17. Schoenmakers, P. J. *J. Chromatogr.* **1984**, *315*, 1.
18. Eckert, C. A.; Ziger, D. H.; Johnston, K. P.; Kim, S. *J. Phys. Chem.* **1986**, *90*, 2738.
19. Springston, S. R.; David, P.; Steger, J.; Novotny, M. *Anal. Chem.* **1986**, *58*, 997.
20. Olesik, S. V.; Steger, J. L.; Kiba, N.; Roth, M.; Novotny, M. V. *J. Chromatogr.* **1987**, *392*, 165.
21. Shim, J.-J.; Johnston, K. P. *AIChE J.* **1989**, *35*, 1097.
22. Shim, J.-J.; Johnston, K. P. *AIChE J.* **1991**, *37*, 607.
23. Shim, J.-J.; Johnston, K. P. *J. Phys. Chem.* **1991**, *95*, 353.
24. Janssen, H.-G.; Snijders, H.; Cramers, C.; Schoenmakers, P. *J. High Resol. Chromatogr.* **1992**, *15*, 458.
25. Roth, M. *J. Chromatogr.* **1993**, *641*, 329.
26. Köhler, U.; Biermanns, P.; Klesper, E. *J. Chromatogr. Sci.* **1994**, *32*, 461.
27. Gönenc, Z. S.; Akman, U.; Sunol, A. K. *J. Chem. Eng. Data* **1995**, *40*, 799.
28. Martin, M.; Blu, G.; Guiochon, G. *J. Chromatogr. Sci.* **1973**, *11*, 641.
29. Bidlingmeyer, B. A.; Hooker, R. P.; Lochmuller, C. H.; Rogers, L. B. *Sep. Sci.* **1969**, *4*, 439.
30. Bidlingmeyer, B. A.; Rogers, L. B. *Sep. Sci.* **1972**, *7*, 131.
31. Chalányová, M.; Macko, T.; Kandrác, J.; Berek, D. *Chromatographia* **1984**, *18*, 668.
32. Berek, D.; Macko, T. *Pure Appl. Chem.* **1989**, *61*, 2041.
33. Tanaka, N.; Yoshimura, T.; Araki, M. *J. Chromatogr.* **1987**, *406*, 247.
34. McGuffin, V. L.; Evans, C. E.; Chen, S.-H. *J. Microcol. Sep.* **1993**, *5*, 3.
35. McGuffin, V. L.; Chen, S.-H. *J. Chromatogr.* **1997**, *762*, 35.
36. Guichon, G.; Sepaniak, M. J. *J. Chromatogr.* **1992**, *606*, 248.
37. Ringo, M. C.; Evans, C. E. *Anal. Chem.* **1997**, *69*, 643.
38. Ringo, M. C.; Evans, C. E. *Anal. Chem.* **1997**, *69*, 4964.
39. Evans, C. E.; Davis, J. *Anal Chim Acta* **1999**, submitted.
40. MacNair, J. E.; Lewis, K. C.; Jorgenson, J. W. *Anal. Chem.* **1997**, *69*, 983.

Chapter 4

Three-Dimensional Stochastic Simulation for the Unified Treatment of Chromatographic and Electrophoretic Separations

Victoria L. McGuffin, Peter E. Krouskop, and Daniel L. Hopkins

Department of Chemistry and Center for Fundamental Materials Research, Michigan State University, East Lansing, MI 48824–1322

A stochastic simulation has been developed to follow the trajectories of individual molecules through the mass transport processes of diffusion, convection by laminar and electroosmotic flow, electrophoretic migration, and surface interaction by an absorption or adsorption mechanism. The molecular zone profile may be examined and characterized at any time or distance during the simulation. This simulation provides a powerful and versatile means to study transport phenomena in chromatography, electrophoresis, and electro-chromatography. In the present work, this simulation is applied to study the kinetic, equilibrium, and fluid dynamic properties of chromatographic systems as a function of the diffusion coefficients in the fluid and surface phases and the interfacial resistance to mass transport.

In recent years, stochastic or Monte Carlo simulation methods have been employed to study a wide variety of systems in separation science, including flow injection analysis (1–3), field–flow fractionation (4,5), electrophoresis (6,7), and chromatography (8–10). An advantage of such methods is that they deal with the behavior of individual molecules, rather than the bulk or average behavior of statistical ensembles. By monitoring the positions or trajectories of these molecules in space and time, a direct and detailed description of physical and chemical phenomena is possible. Stochastic simulations require few, if any, simplifying assumptions and can be designed to include all relevant mass transport processes in a single unified model. These simulation methods can be applied to homogeneous systems or to systems with physical and chemical heterogeneity at the molecular, microscopic, or macroscopic level. The systems may be near to or far from steady-state or equilibrium conditions. Consequently, these simulation methods are a powerful and versatile means to model complex separation systems. Finally, stochastic simulations can provide the connection between molecular-level models, such as *ab initio* quantum mechanics or molecular dynamics calculations, and classical theoretical models or experiments that describe the macroscopic or bulk behavior of the system.

There are several distinctions among the various stochastic simulations that have been developed to date. In these simulations, the fundamental equations of motion for each relevant mass transport process are applied independently to each molecule. The simulations range in complexity from one-dimensional models with finite step size (*11,12*) to three-dimensional models with variable step size (*1–3,7,9,10*) for each transport process. The more rigorous and comprehensive models generally provide the most detailed insight as well as the greatest accuracy and precision. The transport algorithms may be implemented by uniformly incrementing time and calculating the distance traveled (*1–3,7,9,10*), or by incrementing distance and calculating the time required to traverse that distance (*4–6*). Although these approaches may seem equivalent, there is an important difference in the statistical distribution of error. No matter how large the number of molecules or how small the time or distance increment, there is a lower limit to the statistical error (analogous to the well-known Heisenberg uncertainty principle). If time is the incremental variable then its uncertainty or imprecision is controlled and the remaining imprecision is contained in the distance. Conversely, if distance is the incremental variable, then the uncontrolled imprecision is contained in the time (or velocity). Each of these approaches has conditions under which it is advantageous; for example, controlled precision in the distance domain is most beneficial when there is spatial heterogeneity in the system.

Another distinction is that some stochastic simulations are performed by advancing each molecule individually through the separation system (*4–6*), whereas other simulations advance all molecules simultaneously in each time or distance increment (*1–3,7,9,10*). The former approach has the advantage that it can be performed until a given number of molecules have been simulated or until a desired level of precision has been achieved. This approach has the potential to be faster because it utilizes no more than the requisite minimum number of molecules. However, this approach does not easily provide both spatial and temporal distributions of the molecules. Furthermore, this approach is not suitable for modeling transport processes that are dependent upon the local number or concentration of molecules, such as those involving interconversion of the solute by acid/base or complexation reactions, nonideal solute–solute interactions, and nonlinear absorption or adsorption isotherms. For the proper implementation of such processes, all molecules must be advanced simultaneously through the system.

In the present work, a three-dimensional stochastic simulation has been developed for the unified treatment of chromatography, electrophoresis, and electrochromatography. In this simulation, the migration of individual molecules or ions is established through the processes of diffusion and convection by laminar, electroosmotic, and electrophoretic flow. Molecular retention may arise by absorption (partition) into permeable surfaces or by adsorption at solid surfaces that are homogeneous or heterogeneous. The molecular distribution and the corresponding zone profile may be examined and characterized at any specified time or spatial position during the simulation. This simulation has been applied to examine the effects of diffusion in the fluid and surface phases as well as interfacial resistance to mass transport in unified chromatography systems.

Three-Dimensional Stochastic Simulation

The three-dimensional molecular simulation program was written in the FORTRAN 90 programming language and was optimized for execution on an IBM RS/6000 Model 580 computer. This program incorporates algorithms for the processes of diffusion, convection by laminar and electroosmotic flow, electrophoretic migration, and surface interaction by an absorption (partition) or adsorption mechanism, as shown schematically in Figure 1. These processes are applied to each molecule at each time increment (t) until the total simulation time (T) is reached. The simulation may be performed in Cartesian global coordinates, which is most appropriate for separations in planar media, or alternatively in cylindrical global coordinates for

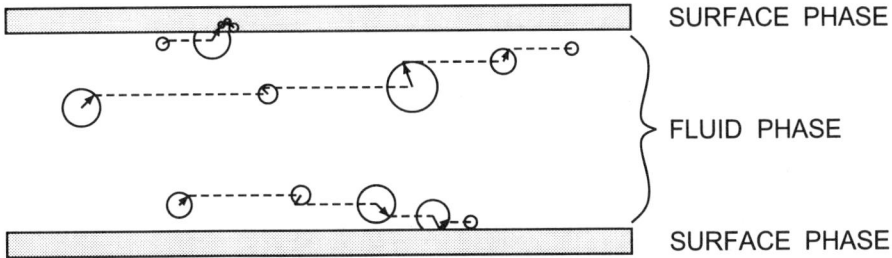

Figure 1. Schematic representation of the trajectories of three molecules during four sequential time increments of the stochastic simulation. Diffusion is illustrated as a sphere of randomly varying radial distance (ρ) with a vector indicating the randomly selected spherical coordinate angles (ϕ,θ). Axial displacement due to laminar or electroosmotic convection is illustrated as a dashed line. Surface interaction is shown for a molecule that is retained by the surface phase (top) and a molecule that is not retained and undergoes an elastic collision at the interface (bottom).

separations in capillary tubes, membranes, or fibers. Because of its mathematical simplicity, the latter case will be described in detail.

Simulation Input. The input parameters required for the simulation may be divided into three general categories, as summarized in Table I. The system parameters describe properties of the fluid and the surface, as well as the spatial dimensions of the separation system to be simulated. The molecular parameters describe attributes of the solute molecules or ions. The values of these parameters may be systematically chosen to characterize the behavior of the system, as in the present study, or may be derived from ab initio calculation or from experiment. On the basis of these input parameters, an array is created that contains the properties and coordinates of each molecule. To initialize the simulation, the molecules are distributed randomly with a

Table I. Input Parameters for the Stochastic Simulation Program.

System Parameters	Symbol
Radius of fluid phase	R_f
Radius of surface phase	R_s
Length of fluid/surface phase	L
Position of injection zone	L_{inj}
Length, variance of injection zone	l_{inj}, σ_{inj}^2
Position of detection zone	L_{det}
Length, variance of detection zone	l_{det}, σ_{det}^2
Zeta potential of surface phase	ζ
Velocity of fluid phase	v_0
pH of fluid phase	pH
pC of complexing agent in fluid phase	pC
Ionic strength of fluid phase	I
Viscosity of fluid phase	η
Dielectric constant of fluid phase	ε
Temperature	T_0
Pressure	P
Voltage	V
Molecular Parameters	**Symbol**
Diffusion coefficient in fluid phase	D_f
Diffusion coefficient in surface phase	D_s
Equilibrium constant for acid/base reaction	K_a
Equilibrium constant for complexation reaction	K_c
Absorption coefficient	K_{abs}
Adsorption energy	E_{ads}
Electrophoretic mobility	μ
Charge	$z_{\pm}$
Computational Parameters	**Symbol**
Number of molecules	N
Time increment	t
Total simulation time	T
Molecular coordinate systems	
Spherical coordinates	ρ, ϕ, θ
Cartesian coordinates	x,y,z
Global coordinate systems	
Cylindrical coordinates	R, Θ, Z
Cartesian coordinates	X,Y,Z

delta, rectangular, or Gaussian profile of specified variance at a specified mean distance in the global coordinate frame. The molecules may be distributed entirely in the fluid phase, entirely in the surface phase, or at equilibrium between the phases.

Simulation Output. The simulation program allows the molecular zone profile to be examined as the distance distribution at specified times or, correspondingly, as the time distribution at specified distances. The statistical moments of the molecular distribution are then calculated in either the distance or time domain (*7,9,10*). For example, the first statistical moment or mean distance $\bar{Z}$ is calculated as

$$\bar{Z} = N^{-1} \sum_{i=1}^{N} Z_i \qquad [1]$$

where Z_i is the axial global coordinate of an individual molecule and N is the total number of molecules. The second statistical moment or variance σ^2 is calculated as

$$\sigma^2 = N^{-1} \sum_{i=1}^{N} (Z_i - \bar{Z})^2 \qquad [2]$$

and the higher statistical moments are calculated in a similar manner. These statistical moments, as well as the chromatographic or electrophoretic figures of merit derived therefrom, are stored in a standard data file at each specified time (or distance). For example, the capacity factor, effective mobility, velocity, plate height, etc. can be calculated since the beginning of the simulation (net average) or since the most recent data file output (local average). Other information such as the number of molecules in the fluid and surface phases, the time spent by each molecule in each phase, and the number of transitions between phases are also recorded in the standard data file.

In addition to these numerical output parameters, the molecular population is summed in discrete segments and then smoothed by Fourier transform methods (*13*) to provide a continuous zone profile for graphical display. Because the molecular distribution may be examined at any time (or distance), these output routines provide an extensive visual and numerical record of transport processes throughout the simulation.

Diffusion. Molecular diffusion is simulated by using a three-dimensional extension of the Einstein–Smoluchowski equation (*14–16*). The radial distance ρ travelled during the time increment t is selected randomly from the following probability distribution

$$P_\rho = \frac{\rho^2}{(4\pi D_{f,s} t)^{1/2}} \exp\left(\frac{-\rho^2}{4 D_{f,s} t}\right) \qquad [3]$$

where $D_{f,s}$ represents the binary diffusion coefficient of the molecule in the fluid or surface phase, as appropriate. This approach provides a variable step size derived from a normal (Gaussian) distribution, where the direction of travel is subsequently randomized through the spherical coordinate angles (ϕ,θ). The coordinate increments in the molecular frame are used to calculate the new molecular position in the global coordinate frame.

To verify the accuracy of the diffusion algorithm, the zone distance and variance for an ensemble of 750 molecules were monitored as a function of the simulation time. These results were compared with classical mass balance models based on the Einstein equation (*14*). Excellent agreement was observed for the range of diffusion coefficients from 10^{-1} to 10^{-10} cm^2 s^{-1}, with average relative errors for the zone distance and variance of 0.81% and 3.67%, respectively (*7,17*).

Convection. Molecular convection in the fluid phase may be induced by means of a pressure or electrical field gradient applied tangential to the surface. The axial distance z travelled by a molecule in time increment t is given by

$$z = vt \qquad [4]$$

For pressure-induced flow under fully developed laminar conditions, the radial velocity profile in the cylindrical global frame is given by the Taylor–Aris equation (*18,19*)

$$v = 2v_0 \left(1 - \frac{R^2}{R_f^2} \right) \qquad v_0 = \frac{R_f^2}{8\eta} \frac{P}{L} \qquad [5a,b]$$

The mean velocity v_0 may be specified as an input parameter or may be calculated from the Hagen–Poiseuille equation (*20–23*), where P is the applied pressure, η is the viscosity of the fluid phase, R_f and L are the radius and length. The coordinate increment in the molecular frame determined from equations [4] and [5a] is used to calculate the new molecular position in the global coordinate frame.

To verify the accuracy of the laminar convection algorithm, the zone distance and variance for an ensemble of 750 molecules were monitored as a function of the simulation time. These results were compared with classical mass balance models based on the Taylor–Aris equation (*18,19*) with both diffusion and resistance to mass transfer in the fluid phase. Excellent agreement was observed for the range of linear velocities from 0.001 to 100 cm s^{-1}, with average relative errors for the zone distance and variance of 0.49% and 2.24%, respectively (*9*).

For electric field-induced flow due to electroosmosis, the radial velocity profile in the cylindrical global frame is given by the Rice–Whitehead equation (*24,25*)

$$v = v_0 \left(1 - \frac{I_0(\kappa R)}{I_0(\kappa R_f)} \right) \qquad v_0 = \frac{\varepsilon \zeta V}{4\pi\eta \ L} \qquad [6a,b]$$

where κ^{-1} is the Debye length, and I_0 is the zero-order modified Bessel function of the first kind. The maximum velocity v_0 may be specified as an input parameter or may be calculated from the Helmholtz–Smoluchowski equation (*26,27*), where ε is the permittivity of the fluid phase, ζ is the zeta potential of the fluid–surface interface, and V is the applied voltage. The coordinate increment in the molecular frame determined from equations [4] and [6a] is used to calculate the new molecular position in the global coordinate frame.

To verify the accuracy of the electroosmotic convection algorithm, the zone distance and variance for an ensemble of 500 molecules were monitored as a function of the simulation time. These results were compared with classical mass balance models based on the analytical solution of the Rice–Whitehead equation by McEldoon and Datta (*28*). Excellent agreement was observed for the range of linear velocities from 0.01 to 1.0 cm s^{-1}, with average relative errors for the zone distance and variance of 0.12% and 4.42%, respectively (*7*).

These convection algorithms may be used individually or in combination to simulate a wide variety of hydrodynamic conditions for chromatography, electrophoresis, or electrochromatography. To demonstrate this ability, the radial flow profiles produced for a constant average velocity of 0.10 cm s^{-1} with varying contributions from laminar and electroosmotic convection are shown in Figure 2. A characteristic feature of these flow profiles is the isosbestic point, which occurs at an ordinate value equal to the average velocity and an abscissa value which divides the fluid phase into equal volumes.

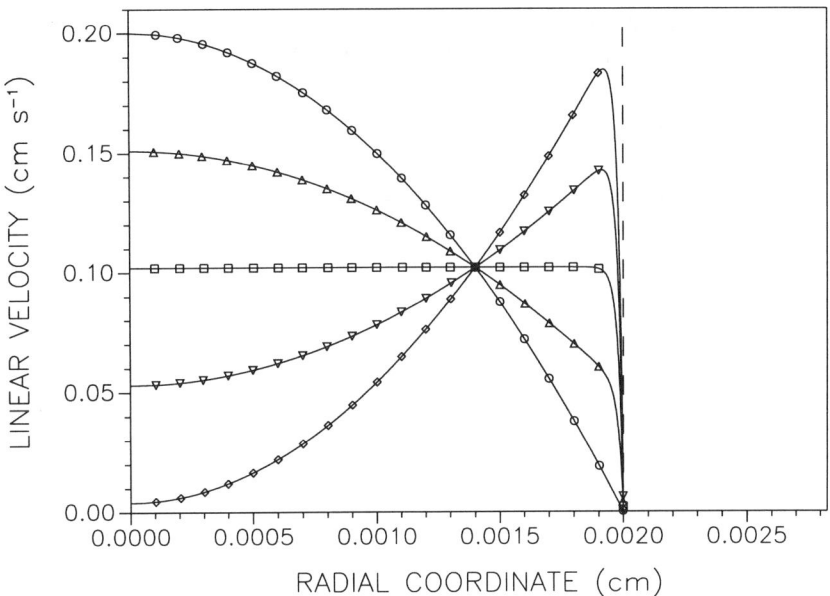

Figure 2. Radial flow profiles for laminar and electroosmotic convection determined by the stochastic simulation. $N = 1.0 \times 10^3$; $t = 1.0 \times 10^{-4}$ s; $R_f = 2.0 \times 10^{-3}$ cm; $R_s = 8.28 \times 10^{-4}$ cm; $v_0 = 0.10$ cm s^{-1} laminar (○), 0.10 cm s^{-1} electroosmotic (□), 0.05 cm s^{-1} laminar and 0.05 cm s^{-1} electroosmotic (△), -0.05 cm s^{-1} laminar and 0.15 cm s^{-1} electroosmotic (▽), -0.10 cm s^{-1} laminar and 0.20 cm s^{-1} electroosmotic (◇), where $\kappa R_f = 100$.

Electrophoretic Migration. For charged molecules under the influence of an applied electric field (*29*), the velocity of electrophoretic migration is given by

$$v = \frac{\mu V}{L} \qquad [7]$$

The electrophoretic mobility μ is corrected by means of the modified Onsager equation (*30*) to the specified ionic strength of the fluid phase.

If the molecule exists as a single species, the mobility is constant. This convection algorithm provides equal displacement of all molecules during each time increment according to equations [4] and [7]. The axial coordinate increment in the molecular frame is used to calculate the new position for each molecule in the global frame.

If the molecule exists as n multiple species in dynamic equilibrium (e.g., phosphate may exist as H_3PO_4, $H_2PO_4^-$, HPO_4^{2-}, or PO_4^{3-}), the mobility of an individual molecule is determined from statistical probability at each time increment. The fraction α_i of each species i is calculated from the appropriate equilibrium constants for acid/base or complexation reactions, which are corrected for ionic strength by means of the Davies equation (*31,32*). The identity of a molecule is determined by selecting a random number, ξ, between zero and one to establish the value of i that satisfies the relationship

$$1 - \sum_{j=1}^{n-i} \alpha_{n-j} < \xi \leq \sum_{j=1}^{i+1} \alpha_{j-1} \qquad [8]$$

The molecule is then assigned the mobility μ_i corresponding to species i during that time increment and its electrophoretic migration is calculated via equations [4] and [7]. The resulting migration of the zone is similar, but not identical, to that for a single species whose average mobility is given by

$$\mu = \sum_{i=0}^{n} \alpha_i \, \mu_i \qquad [9]$$

To verify the accuracy of the electrophoretic migration algorithms corresponding to a single species and n multiple species, the zone distance and variance for an ensemble of 500 molecules were monitored as a function of the simulation time. These results were compared with classical models. Excellent agreement was observed for single species with positive and negative electrophoretic mobilities in the range from $+10^{-3}$ to -10^{-3} cm^2 V^{-1} s^{-1}, with average relative errors for the zone distance and variance of 0.04% and 2.67%, respectively (*7*). The agreement is similarly good for multiple species, with average relative errors for the zone distance and variance of 0.01% and 3.38%, respectively, for phosphate at pH values from 3.0 to 9.0 (*7*).

Surface Interaction. Molecular interaction with a stationary surface is simulated as an absorption process if the surface is permeable (e.g., thin polymer film or chemically bonded organic ligands) or as an adsorption process if the surface is solid (e.g., silica or alumina).

For the absorption process, the probability of transport between the fluid and surface phases is given by

$$\begin{cases} P_{fs} = aK_{abs}(D_s/D_f)^{1/2} \\ P_{sf} = a \end{cases} \text{ or } \begin{cases} P_{fs} = a \\ P_{sf} = aK_{abs}^{-1}(D_f/D_s)^{1/2} \end{cases} \quad [10a,b]$$

where K_{abs} is the absorption coefficient and the constant a represents the fraction of effective collisions with the interface, which is equal to unity when there is no barrier to transport (diffusion-limited case). When a molecule in the fluid phase encounters the fluid–surface interface during the simulation, a random number between zero and one is selected. If the selected number is less than or equal to the probability P_{fs} given by equation [10a,b], the molecule will be transferred to the surface phase. Otherwise, the molecule will remain in the fluid phase and will undergo an elastic collision at the interface. A similar routine is performed when a molecule in the surface phase encounters the interface, except that the random number is compared with the probability P_{sf} given in equation [10a,b]. Finally, when a molecule in the fluid or surface phase encounters a physical boundary of the system, an elastic collision is performed.

To verify the accuracy of the absorption algorithm, the zone distance and variance for an ensemble of 750 molecules were monitored as a function of the simulation time. These results were compared with classical mass balance models based on the extended Golay equation (*33*) including both diffusion and resistance to mass transfer in the fluid and surface phases. Excellent agreement was observed for the range of distribution coefficients from 0.01 to 100.0, with average relative errors for the zone distance and variance of 0.55% and 4.02%, respectively (*9*).

For the adsorption process, the surface is considered to be a uniform distribution of localized lattice sites, each of equal area and interaction energy with the molecule. The molar adsorption energy E_{ads} is related to the mean time for desorption τ_{ads} in the following manner

$$\tau_{ads} = \tau_0 \exp\left(\frac{E_{ads}}{k_A k_B T_0}\right) \quad [11]$$

where τ_0 is the vibrational period, typically $10^{-12} - 10^{-13}$ s, T_0 is the temperature, k_B is the Boltzmann constant and k_A is the Avogadro number. If a molecule encounters the surface, the probability for adsorption P_{fs} is equal to unity if the site is vacant and zero if it is occupied. The desorption time of the molecule is then randomly selected from an exponential distribution based on the mean desorption time τ_{ads} in equation [11].

The absorption and adsorption algorithms may be used for homogeneous surfaces, as described above, or for heterogeneous surfaces with a fractional coverage of two or more types of surface sites (*34*). In this case, a random number is selected to establish the identity of the individual surface site by comparison with the fractional coverage. A second random number is then selected to determine whether the molecule is transferred to the surface phase for the absorption mechanism or to determine the desorption time for the adsorption mechanism. These algorithms may be used individually or in combination to describe a wide variety of retention mechanisms for chromatography, electrophoresis, and electrochromatography.

Selected Applications of the Stochastic Simulation

Some applications have been selected to illustrate the capabilities and versatility of the stochastic simulation approach. The kinetic and equilibrium behavior are characterized for a model chromatographic system with a simple absorption mechanism under diffusion-limited conditions. In the first series of simulations, the behavior is examined as a function of the diffusion coefficient in the fluid phase. In the second series of simulations, the effect of the diffusion coefficient in the surface phase is similarly explored. Finally, the influence of interfacial resistance to mass transport between the fluid and surface phases is examined.

For each of these cases, the kinetic behavior of the system is elucidated by monitoring the number of molecules in the fluid phase as a function of the simulation time. These data are analyzed by means of nonlinear regression to the following equation

$$\frac{N_f}{N} = \frac{k_{sf} + k_{fs} \exp(-(k_{fs} + k_{sf})T)}{k_{fs} + k_{sf}} \qquad [12]$$

in order to determine the pseudo first-order rate constants for transport between the fluid and surface phases. By using this approach, the rate constants k_{fs} and k_{sf} can typically be determined with ±0.49% relative standard deviation and the ratio of the rate constants k_{fs}/k_{sf} with ±0.70% relative standard deviation and ±2.25% relative error (10,35). The characteristic time τ is given by

$$\tau = \frac{1}{k_{fs} + k_{sf}} \qquad [13]$$

The equilibrium behavior of the system is elucidated by monitoring the number of molecules in the fluid and surface phases at equilibrium, $\tilde{N}_f$ and $\tilde{N}_s$, respectively. The ratio of the number of molecules can typically be determined with ±0.29% relative standard deviation and ±0.39% relative error (10). The kinetic and equilibrium descriptions of the system are related in the following manner:

$$\frac{k_{fs}}{k_{sf}} = \frac{\tilde{N}_s}{\tilde{N}_f} = \frac{K_{abs} V_s}{V_f} \qquad [14]$$

where the volumes of the fluid and surface phases are given as $V_f = \pi R_f^2 L$ and $V_s = \pi (R_s^2 + 2 R_s R_f) L$, respectively.

Finally, for each case, the hydrodynamic behavior of the system is elucidated under laminar flow conditions. In the presence of flow, the characteristic time τ will influence the appearance of the solute zones. If τ is sufficiently small, the system will be nearly at equilibrium and the zone profile will be a symmetric Gaussian distribution. Under these conditions, the profile will be well described by classical equations of mass balance based on the equilibrium–dispersive model, such as the Golay equation (33). As τ increases, however, the system may depart from equilibrium and the zone profile may become highly asymmetric. As a measure of the degree of departure from equilibrium for convective systems, we may define a unitless kinetic parameter P as

$$P = \frac{\tau}{T} = \frac{\tau v_0}{\overline{Z}} \qquad [15]$$

where T is the time, $\overline{Z}$ is the mean distance travelled, and v_0 is the mean linear velocity. This parameter directly reflects the sources of kinetic stress that are placed on the system and will approach a limiting value of zero for a system that is at equilibrium. For each of the cases outlined above, the solute zone profiles are simulated at fixed times from 0 to 30 s. The mean zone distance and variance are determined by means of equations [1] and [2] as a function of the simulation time and are compared with classical theoretical models (33).

Effect of the Diffusion Coefficient in the Fluid Phase. The stochastic simulation method is especially well suited for the unified treatment of chromatographic separations. To illustrate this capability, we have simulated systems that are representative of gas, dense gas, supercritical fluid, enhanced fluidity liquid, and liquid chromatography. The kinetic behavior of these systems is illustrated in Figure 3. The gas, dense gas, and supercritical fluid are compared as fluid phases in an open-tubular column with radius R_f of 50 μm and surface phase R_s of 0.25 μm, resulting in a volumetric phase ratio V_f/V_s of 100. As shown in Figure 3A, the kinetic behavior of the system is nearly indistinguishable, despite the significant change in diffusion coefficients D_f from 1.0×10^{-1} to 1.0×10^{-3} cm^2 s^{-1} for these fluid phases. This observation is confirmed by the rate constants k_{fs} and k_{sf}, which were determined by nonlinear regression of the simulation data to equation [12] and are summarized in Table II. There is a small but statistically significant decrease in both k_{fs} and k_{sf} with decreasing diffusion coefficient in the fluid phase. The overall transport rate, as represented by the characteristic time τ in equation [13], is controlled by the rate constant for transport from the surface to fluid phase (k_{sf}) which is one-hundred-fold larger than that from fluid to surface phase (k_{fs}). It is evident that the characteristic time τ is very small (~10^{-5} s), which indicates that equilibrium is rapidly achieved within this system for all of the fluid phases. Finally, the ratio of the rate constants k_{fs}/k_{sf} and the ratio of the number of molecules at equilibrium $\tilde{N}_s/\tilde{N}_f$ are in good agreement with the theoretically predicted value of $K_{abs} V_f/V_s = 0.01$ given by equation [14].

The supercritical fluid, enhanced fluidity liquid, and liquid are compared as fluid phases in an open-tubular column with radius R_f of 20 μm and surface phase R_s of 8.28 μm, resulting in a volumetric phase ratio V_f/V_s of 1.0. As shown in Figure 3B, the kinetic behavior of the system is somewhat more distinguishable for these fluid phases with diffusion coefficients D_f from 1.0×10^{-3} to 1.0×10^{-5} cm^2 s^{-1}. The rate constants derived from these data are summarized in Table II. The rate constants k_{fs} and k_{sf} are approximately equal, as expected from equation [14], and are several orders of magnitude smaller than those determined for the system described above. The characteristic time τ is significantly larger (~10^{-2} s), so that equilibrium is much more slowly achieved than in the system above.

Table II. Effect of Diffusion Coefficient in the Fluid Phase on Rate Constants.

D_f (cm^2 s^{-1})	k_{fs} (s^{-1})	k_{sf} (s^{-1})	τ (s)	k_{fs}/k_{sf}	$\tilde{N}_s/\tilde{N}_f$
1.0×10^{-1} [a]	442	44000	2.25×10^{-5}	0.01006	0.01021
1.0×10^{-2} [a]	407	40700	2.44×10^{-5}	0.01000	0.00996
1.0×10^{-3} [a]	367	37200	2.66×10^{-5}	0.00986	0.01000
1.0×10^{-3} [b]	36.6	37.0	0.014	0.989	0.995
1.0×10^{-4} [b]	30.5	30.7	0.016	0.993	1.003
1.0×10^{-5} [b]	13.74	13.87	0.036	0.991	0.997

[a] Simulation conditions: $N = 1.0 \times 10^5$; $t = 1.0 \times 10^{-7}$ s; $T = 20 \tau$; $R_f = 5.0 \times 10^{-3}$ cm; $R_s = 2.5 \times 10^{-5}$ cm; $D_s = 1.0 \times 10^{-5}$ cm^2 s^{-1}; $K_{abs} = 1.0$.
[b] Simulation conditions: $N = 1.0 \times 10^4$; $t = 1.0 \times 10^{-5}$ s; $T = 20 \tau$; $R_f = 2.0 \times 10^{-3}$ cm; $R_s = 8.28 \times 10^{-4}$ cm; $D_s = 1.0 \times 10^{-5}$ cm^2 s^{-1}; $K_{abs} = 1.0$.

The liquid chromatography case may be used as a representative example to illustrate other kinetic and equilibrium information that can be derived from the stochastic simulation approach. During the kinetic process shown in Figure 3, the system gradually evolves from the initial condition, where all molecules are uniformly distributed in the fluid phase, to the equilibrium condition, where $\tilde{N}_f = N (K_{abs} V_s/V_f)/(1 + K_{abs} V_s/V_f)$ molecules are uniformly distributed in the

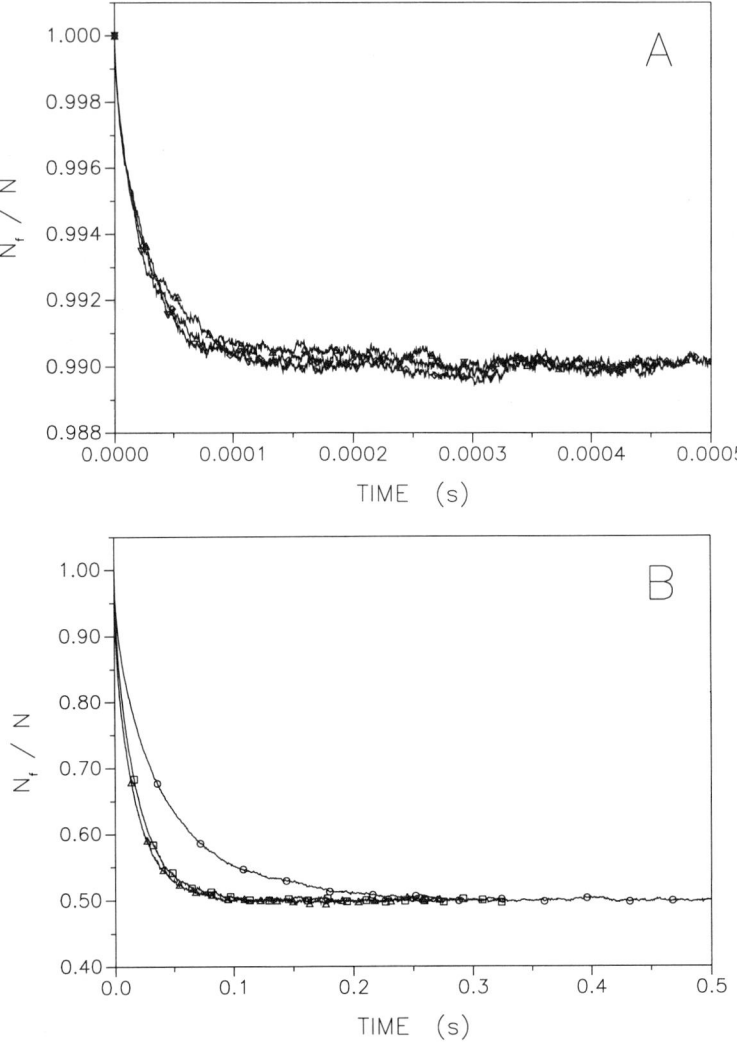

Figure 3. Kinetic evolution of the absorption process with varying diffusion coefficients in the fluid phase representative of (A) gas, dense gas, and supercritical fluid chromatography, (B) supercritical fluid, enhanced fluidity liquid, and liquid chromatography. Simulation conditions: (A) $N = 1.0 \times 10^5$; $t = 1.0 \times 10^{-7}$ s; $T = 20$ τ; $R_f = 5.0 \times 10^{-3}$ cm; $R_s = 2.5 \times 10^{-5}$ cm; $D_f = 1.0 \times 10^{-1}$ cm² s⁻¹ ($\triangledown$), 1.0×10^{-2} cm² s⁻¹ ($\diamondsuit$), 1.0×10^{-3} cm² s⁻¹ ($\triangle$); $D_s = 1.0 \times 10^{-5}$ cm² s⁻¹; $K_{abs} = 1.0$. (B) $N = 1.0 \times 10^4$; $t = 1.0 \times 10^{-5}$ s; $T = 20$ τ; $R_f = 2.0 \times 10^{-3}$ cm; $R_s = 8.28 \times 10^{-4}$ cm; $D_f = 1.0 \times 10^{-3}$ cm² s⁻¹ ($\triangle$), 1.0×10^{-4} cm² s⁻¹ ($\square$), 1.0×10^{-5} cm² s⁻¹ ($\bigcirc$); $D_s = 1.0 \times 10^{-5}$ cm² s⁻¹; $K_{abs} = 1.0$.

fluid phase and $\tilde{N}_s = N/(1 + K_{abs} V_s/V_f)$ molecules are uniformly distributed in the surface phase. The radial distribution of solute molecules is illustrated in Figure 4 at times corresponding to 0.0 τ, 0.1 τ, 0.2 τ, 0.5 τ, 1.0 τ, 2.0 τ, and 5.0 τ, where τ is 0.036 s for the liquid chromatography system. At these times, the system has achieved 0%, 9.5%, 18.1%, 39.4%, 63.2%, 86.5%, and 99.3%, respectively, of the molecular distribution at equilibrium. Once the system has achieved equilibrium (T > 20 τ), we may further characterize the transport between the fluid and surface phases. The residence time distribution for a single sojourn of a molecule in the fluid and surface phases is shown in Figure 5. From this distribution, the average residence time is 1.9×10^{-3} s and the standard deviation is 1.1×10^{-3} s. However the most probable residence time is one time constant, with approximately 30% of the molecules residing just 1.0×10^{-5} s before transferring to the opposite phase. The molecules are transferred between phases an average of 284 times per second at equilibrium. The standard deviation is 76 s^{-1}, which suggests that 95% of the molecules are transferred between phases from 135 to 433 times per second.

The effect of the diffusion coefficient in the fluid phase on the fluid dynamic behavior of the system has also been examined. The solute zone profiles for supercritical fluid, enhanced fluidity liquid, and liquid chromatography at a mean linear velocity of 0.10 cm s^{-1} are shown in Figure 6. Since the simulation time T is much greater than the characteristic time τ in all cases, the system is nearly at equilibrium according to equation [15] and the zone profiles are symmetric. The statistical moments of these zone profiles calculated by means of equations [1] and [2] are shown as a function of the simulation time in Figure 7. The mean distance coincides with the theoretically expected value of $Z = v_0 T/(1 + K_{abs} V_s/V_f)$ for all fluid phases. The variance also agrees well with the extended Golay equation (33), which includes dispersion arising from axial diffusion in the fluid and surface phases and resistance to mass transfer in the fluid and surface phases. The variance for enhanced fluidity liquid chromatography is the smallest because the selected velocity of 0.10 cm s^{-1} is near the optimum value of 0.11 cm s^{-1} for this system. For liquid chromatography, the selected velocity is greater than the optimum value of 0.022 cm s^{-1} and the variance has a correspondingly larger contribution from resistance to mass transfer in the fluid phase. For supercritical fluid chromatography, the selected velocity is less than the optimum value of 0.41 cm s^{-1} and the variance has a correspondingly larger contribution from axial diffusion in the fluid phase. In all cases, however, there is excellent agreement between the simulation results and the extended Golay equation (33) because the system is nearly at equilibrium.

The liquid chromatography case may again be used as a representative example to illustrate other hydrodynamic information that can be derived from the stochastic simulation approach. Consider the final zone profile in Figure 6C, where each molecule has travelled for a fixed total time of 30.0 s. Of this total time, each molecule has spent some time in the fluid phase and the remainder in the surface phase. The residence time distribution in each phase is shown in Figure 8. It is evident that the residence time distribution in the fluid phase is symmetric and the mean value of 15.0 s coincides with the theoretically expected value of $T_f = T/(1 + K_{abs} V_s/V_f) = 15.0$ s. Similarly, the mean residence time in the surface phase of 15.0 s coincides with the expected value of $T_s = T (K_{abs} V_s/V_f)/(1 + K_{abs} V_s/V_f) = 15.0$ s. The standard deviation of these residence time distributions is 0.77 s, which suggests that 95% of the molecules have spent from 13.5 to 16.5 s in each phase. This information can also be used to calculate the capacity factor (k) for each molecule as T_s/T_f. These calculations suggest that the mean capacity factor is 1.00, the standard deviation is 0.10, and 95% of the molecules have capacity factors ranging from 0.80 to 1.20. Finally, it is instructive to graph the residence time in each phase *versus* the distance travelled for each molecule, as shown in Figure 9. This graph confirms that molecules at the front of the zone have spent the greatest time in the fluid phase and the least time in the surface phase, whereas the converse is true for molecules at the

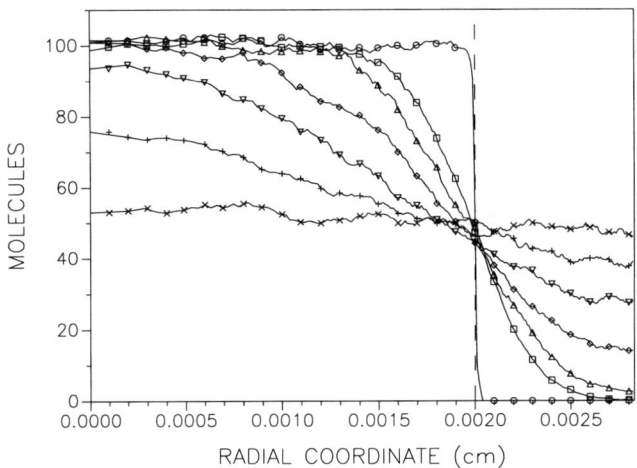

Figure 4. Radial solute distribution profiles during the kinetic evolution of chromatographic systems. Simulation conditions: $N = 1.0 \times 10^4$; $t = 1.0 \times 10^{-5}$ s; $T = 0.0\ \tau$ (○), $0.1\ \tau$ (□), $0.2\ \tau$ (△), $0.5\ \tau$ (◇), $1.0\ \tau$ (▽), $2.0\ \tau$ (+), $5.0\ \tau$ (×), where $\tau = 0.036$ s; $R_f = 2.0 \times 10^{-3}$ cm; $R_s = 8.28 \times 10^{-4}$ cm; $D_f = 1.0 \times 10^{-5}$ cm^2 s^{-1}; $D_s = 1.0 \times 10^{-5}$ cm^2 s^{-1}; $K_{abs} = 1.0$.

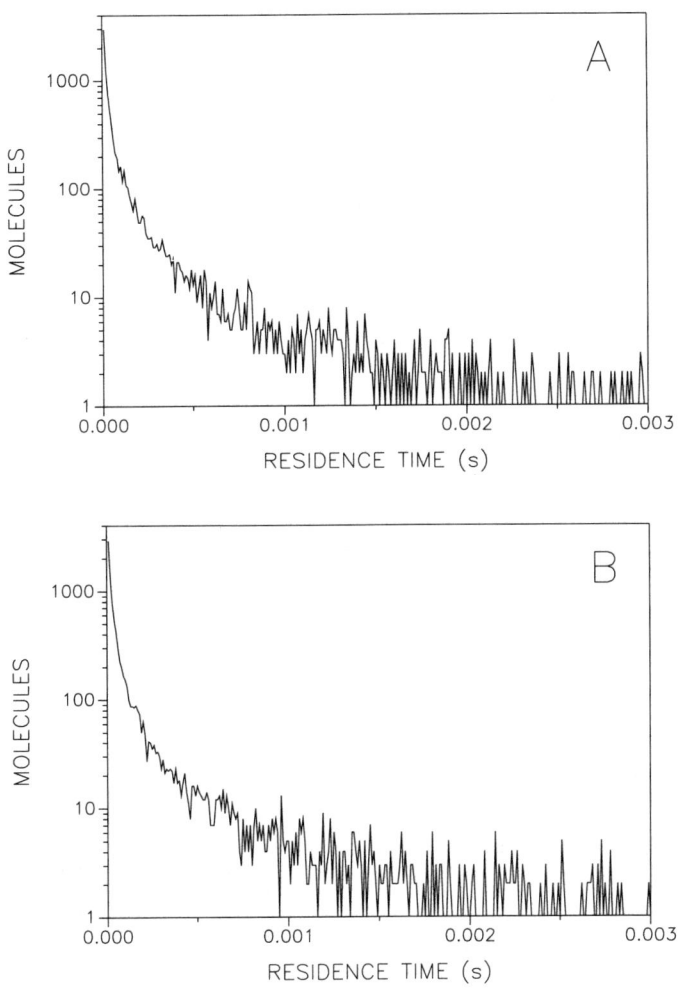

Figure 5. Residence time distribution for a single sojourn in the fluid phase (A) and surface phase (B) under equilibrium conditions. Simulation conditions: $T > 20\,\tau$, where $\tau = 0.036$ s; other conditions as given in Figure 4.

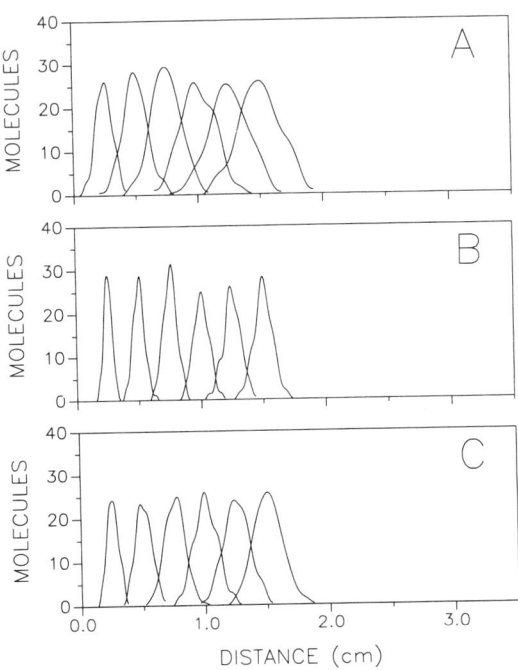

Figure 6. Evolution of the solute zone profile with varying diffusion coefficients in the fluid phase representative of (A) supercritical fluid, (B) enhanced fluidity liquid, and (C) liquid chromatography. Simulation conditions: $N = 1.0 \times 10^3$; $t = 1.0 \times 10^{-5}$ s; $T = 5, 10, 15, 20, 25, 30$ s; $R_f = 2.0 \times 10^{-3}$ cm; $R_s = 8.28 \times 10^{-4}$ cm; $D_f = 1.0 \times 10^{-3}$ cm² s⁻¹ (A), 1.0×10^{-4} cm² s⁻¹ (B), 1.0×10^{-5} cm² s⁻¹ (C); $D_s = 1.0 \times 10^{-5}$ cm² s⁻¹; $K_{abs} = 1.0$, $v_0 = 0.10$ cm s⁻¹.

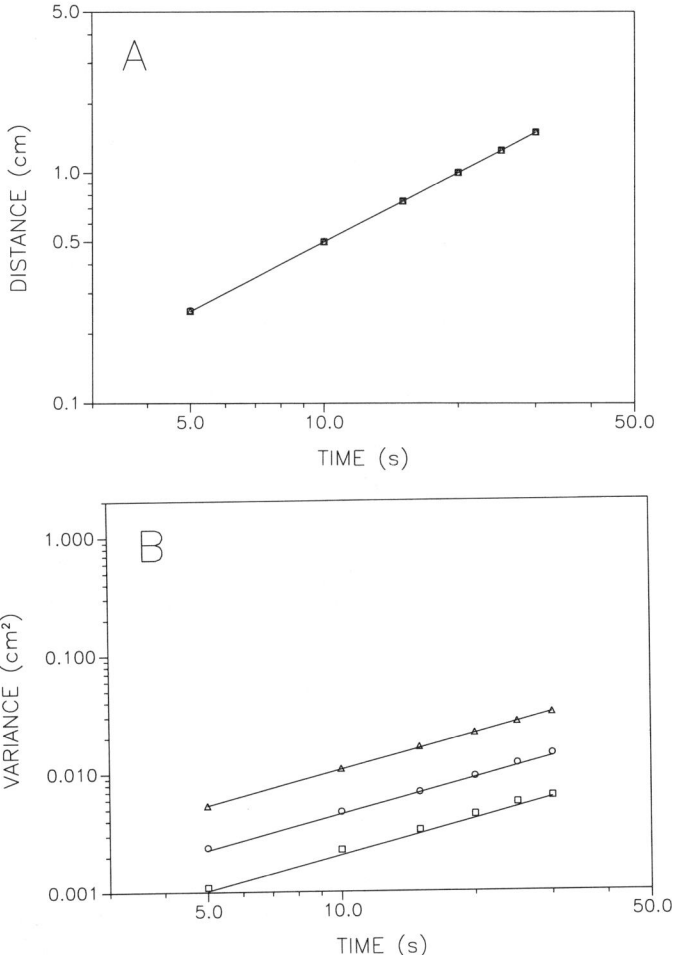

Figure 7. Mean distance (A) and variance (B) of the solute zone profile with varying diffusion coefficients in the fluid phase. Simulation conditions: D_f = 1.0×10^{-3} cm^2 s^{-1} (△), 1.0×10^{-4} cm^2 s^{-1} (□), 1.0×10^{-5} cm^2 s^{-1} (○); other conditions as given in Figure 6C. (———) Theory according to the extended Golay equation (*33*).

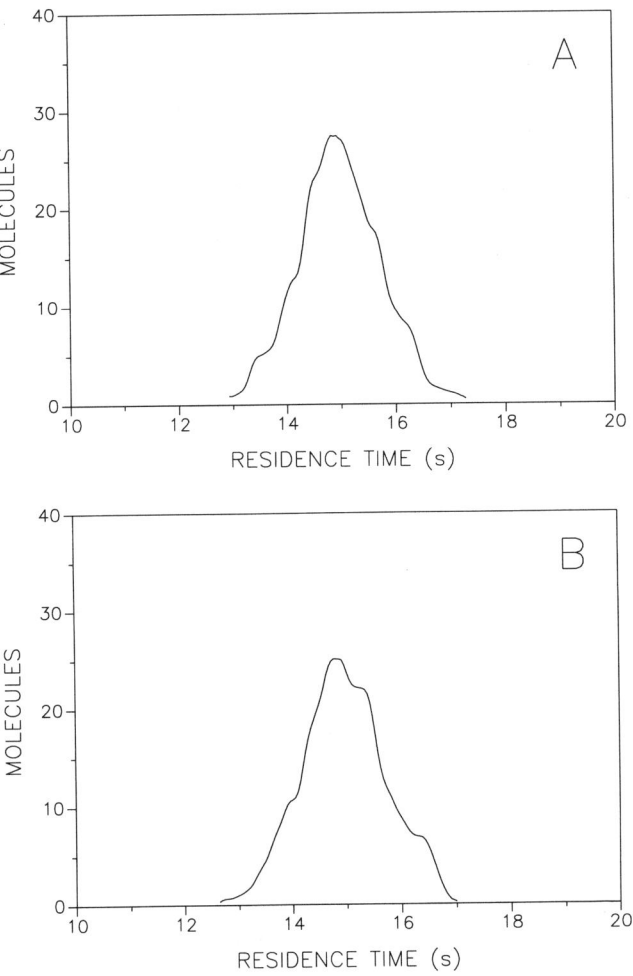

Figure 8. Residence time distribution for total time spent in the fluid phase (A) and surface phase (B). Simulation conditions as given in Figure 6C.

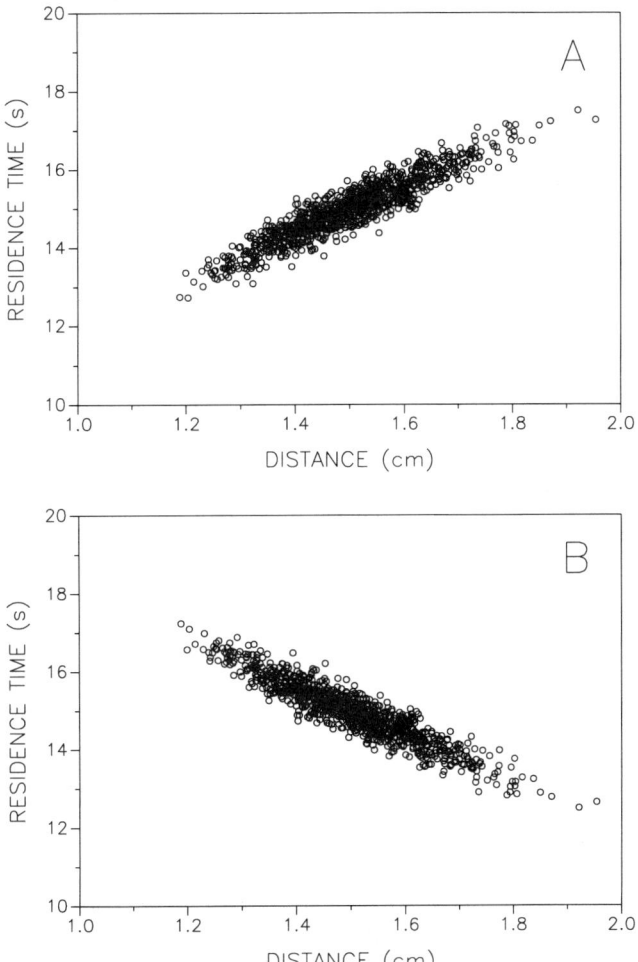

Figure 9. Relationship between total time spent in fluid phase (A) and surface phase (B) and the distance travelled by individual molecules. Simulation conditions as given in Figure 6C.

rear of the zone. The relationship between residence time and distance travelled appears to be linear with slopes of 6.00 and −6.00 s cm^{-1} for the fluid and surface phases, respectively. This slope is a direct measure of the extent of deviation from equilibrium across the solute zone. The steeper the slope, which is related to the variables in equation [15] such as characteristic time τ, velocity, and distance travelled, the greater is the nonequilibrium. Moreover, the broader the solute zone for a given slope, the greater is the nonequilibrium. The data in Figure 9 also suggest that there is significant variation in the behavior of individual molecules. For example, molecules at the center of the zone that have travelled the mean distance of 1.5 cm may have spent from 14.4 to 15.6 s in the fluid and surface phases and may have capacity factors ranging from 0.91 to 1.07 (95% confidence level). These descriptions and characterizations of molecular behavior appear to be typical of solute zones in systems that are nearly at equilibrium.

Effect of the Diffusion Coefficient in the Surface Phase. Using the liquid chromatography case as a representative example, the diffusion coefficient in the surface phase was varied from 1.0×10^{-5} to 1.0×10^{-8} cm^2 s^{-1}. The kinetic behavior of the system is summarized in Figure 10 and Table III. It is apparent from the rate constants and the characteristic time τ that the kinetic behavior is reasonably rapid for the diffusion coefficient of 1.0×10^{-5} cm^2 s^{-1}, slightly slower for 1.0×10^{-6} cm^2 s^{-1}, and significantly slower for 1.0×10^{-7} and 1.0×10^{-8} cm^2 s^{-1}.

Table III. Effect of Diffusion Coefficient in the Surface Phase on Rate Constants.[a]

D_s (cm^2 s^{-1})	k_{fs} (s^{-1})	k_{sf} (s^{-1})	τ (s)	k_{fs}/k_{sf}	$\tilde{N}_s/\tilde{N}_f$
1.0×10^{-5}	13.74	13.87	0.036	0.991	0.997
1.0×10^{-6}	3.076	3.102	0.162	0.991	1.006
1.0×10^{-7}	0.353	0.357	1.41	0.989	0.999
1.0×10^{-8}	0.038	0.039	13.05	0.986	0.999

[a] Simulation conditions: $N = 1.0 \times 10^4$; $t = 1.0 \times 10^{-5}$ s; $T = 20\ \tau$; $R_f = 2.0 \times 10^{-3}$ cm; $R_s = 8.28 \times 10^{-4}$ cm; $D_f = 1.0 \times 10^{-5}$ cm^2 s^{-1}; $K_{abs} = 1.0$.

In order to understand the effect of the diffusion coefficients in the fluid and surface phases on the rate constants, it is helpful to represent the data in Tables II and III graphically. As shown in Figure 11, the rate constants are intrinsically related to the reduced diffusion coefficient $D = D_f D_s/(D_f + D_s)$ for the system (10). When the diffusion coefficients are comparable in magnitude, they both influence the kinetic behavior of the system. However, when one diffusion coefficient is significantly smaller than the other, it serves to limit the overall rate of transport in the system. The consequences of this dependence on the reduced diffusion coefficient were observed in Figure 3 and Table II above. Because the diffusion coefficient in the surface phase was significantly smaller (1.0×10^{-5} cm^2 s^{-1}), the fluid phases of gas, dense gas, and supercritical fluid had little effect on the kinetic behavior. The practical implications of this statement are clear: In the development of unified chromatography, we cannot simply be concerned with the properties of the fluid phase. We must also concomitantly increase the diffusion coefficients in the surface phase so that they are maintained within approximately two orders of magnitude of those in the fluid phase in order to derive the full benefits of the improved kinetic behavior.

The effect of the diffusion coefficient in the surface phase on the fluid dynamic behavior of the system has also been examined. The solute zone profiles obtained at a mean linear velocity of 0.10 cm s^{-1} are shown in Figure 12. For the diffusion

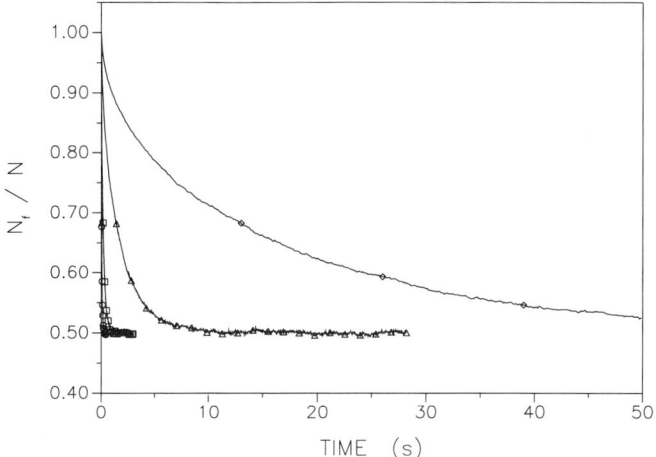

Figure 10. Kinetic evolution of the absorption process with varying diffusion coefficients in the surface phase for liquid chromatography. Simulation conditions: $D_f = 1.0 \times 10^{-5}$ cm^2 s^{-1}; $D_s = 1.0 \times 10^{-5}$ cm^2 s^{-1} (○), 1.0×10^{-6} cm^2 s^{-1} (□), 1.0×10^{-7} cm^2 s^{-1} (△), 1.0×10^{-8} cm^2 s^{-1} (◇); other conditions as given in Figure 3B.

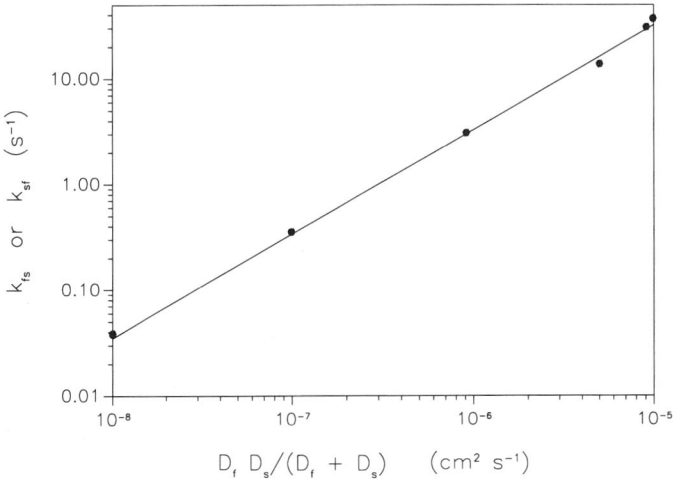

Figure 11. Effect of the reduced diffusion coefficient on rate constants k_{fs} (○) and k_{sf} (●). Simulation conditions as given in Tables II and III.

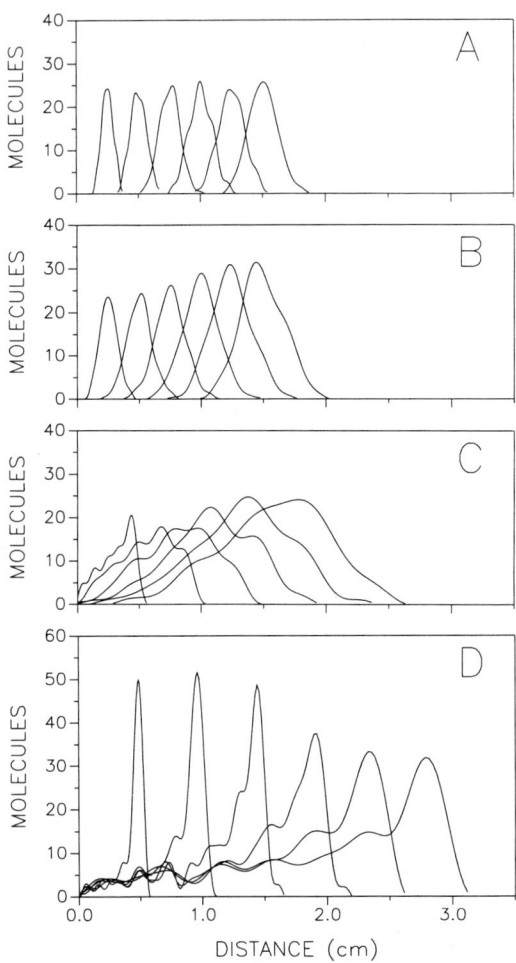

Figure 12. Evolution of the solute zone profile with varying diffusion coefficients in the surface phase for liquid chromatography. Simulation conditions: $D_f = 1.0 \times 10^{-5}$ cm^2 s^{-1}; $D_s = 1.0 \times 10^{-5}$ cm^2 s^{-1} (A), 1.0×10^{-6} cm^2 s^{-1} (B), 1.0×10^{-7} cm^2 s^{-1} (C), 1.0×10^{-8} cm^2 s^{-1} (D); other conditions as given in Figure 6C.

coefficient of 1.0×10^{-5} cm^2 s^{-1}, the ratio of the characteristic time τ to the simulation time T ranges from 0.007 to 0.001 for simulation times from 5 to 30 s, respectively. Consequently, this system is nearly at equilibrium according to equation [15] and all of the zone profiles are symmetric. Similarly for the diffusion coefficient of 1.0×10^{-6} cm^2 s^{-1}, the ratio of τ/T ranges from 0.032 to 0.005 and all of the solute zones appear to be symmetric. However, for the diffusion coefficient of 1.0×10^{-7} cm^2 s^{-1}, the ratio of τ/T is significantly larger and ranges from 0.28 to 0.047. The zone profiles initially deviate from equilibrium and are asymmetric but progressively become more symmetric. The deviations from equilibrium are even more problematic for the diffusion coefficient of 1.0×10^{-8} cm^2 s^{-1}, where the ratio of τ/T ranges from 2.61 to 0.44 and all of the profiles are markedly asymmetric.

The statistical moments of the solute zone profiles are shown as a function of the simulation time in Figure 13. For the diffusion coefficient of 1.0×10^{-5} cm^2 s^{-1}, the mean distance and variance agree well with the theoretically expected values from the extended Golay equation (*33*). However, as the diffusion coefficient in the surface phase decreases and the system deviates from equilibrium behavior, the mean zone distance increases because the molecules have proportionately greater residence time in the fluid phase. The variance also increases because the resistance to mass transfer in the surface phase has an additional contribution from slow kinetics. When the molecular zone begins migration from the initial nonequilibrium state, the variance increases as the square of the simulation time (as evident for the diffusion coefficient of 1.0×10^{-8} cm^2 s^{-1}). As the system gradually evolves toward the steady state, the variance progressively changes until it increases linearly with the simulation time (as evident for the diffusion coefficient of 1.0×10^{-7} cm^2 s^{-1}). As shown in Figure 13, the slow kinetics of the system influence the time required for the onset of steady-state conditions and the variance incurred during this transition as well as the variance per unit time (or length) once steady-state conditions have been achieved. The extended Golay equation (*33*) and other equilibrium-dispersive models cannot be used to predict the behavior of such systems.

Effect of Interfacial Resistance to Mass Transport. It is possible to extend this simulation approach to consider the situation where there is some resistance to mass transport at the interface and all collisions are not sufficiently energetic to overcome this barrier. To represent this situation, the constant *a* in the probability expressions of equation [10a,b] was varied from 1.0 to 0.001 for the case of liquid chromatography. The kinetic behavior of the system is summarized in Figure 14 and Table IV. It is apparent from the rate constants and the characteristic time τ that the kinetic behavior is reasonably rapid for $a = 1.0$, slightly slower for $a = 0.5$, and significantly slower for smaller values of *a*.

Table IV. Effect of Interfacial Resistance to Mass Transport on Rate Constants.[a]

a	k_{fs} (s^{-1})	k_{sf} (s^{-1})	τ (s)	k_{fs}/k_{sf}	$\tilde{N}_s/\tilde{N}_f$
1.0	13.74	13.87	0.036	0.991	0.997
0.5	11.60	11.73	0.043	0.989	0.997
0.1	2.509	2.509	0.199	1.000	1.001
0.05	1.301	1.302	0.384	0.999	0.998
0.01	0.273	0.272	1.84	1.004	1.008
0.005	0.137	0.137	3.65	1.001	1.002
0.001	0.028	0.028	18.02	0.992	1.005

[a] Simulation conditions: $N = 1.0\times10^4$; $t = 1.0\times10^{-5}$ s; $T = 20\ \tau$; $R_f = 2.0\times10^{-3}$ cm; $R_s = 8.28\times10^{-4}$ cm; $D_f = 1.0\times10^{-5}$ cm^2 s^{-1}; $D_s = 1.0\times10^{-5}$ cm^2 s^{-1}; $K_{abs} = 1.0$.

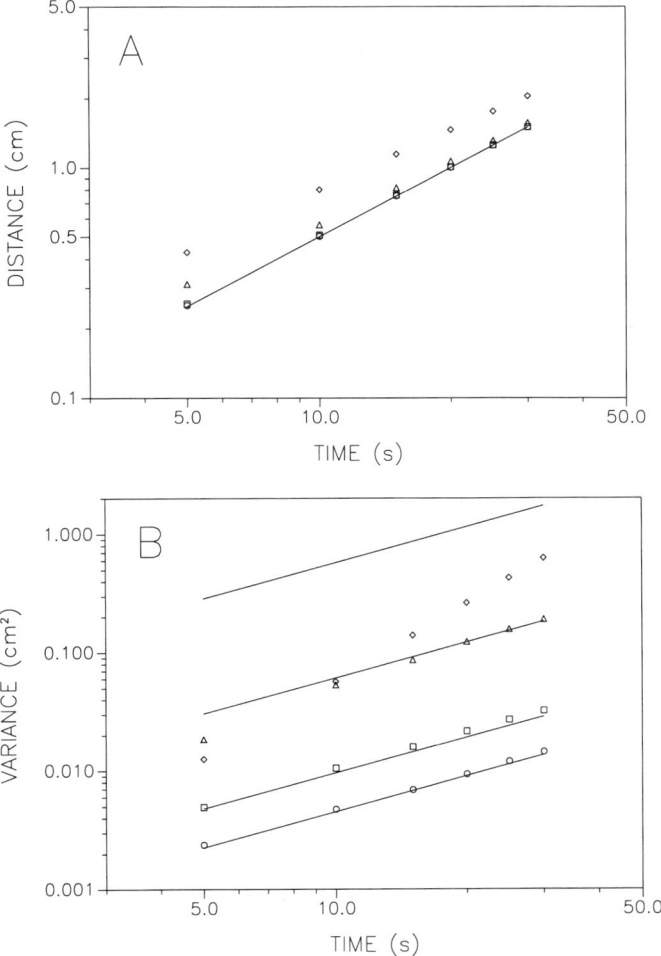

Figure 13. Mean distance (A) and variance (B) of the solute zone profile with varying diffusion coefficients in the surface phase. Simulation conditions: $D_f = 1.0 \times 10^{-5}$ cm^2 s^{-1}; $D_s = 1.0 \times 10^{-5}$ cm^2 s^{-1} (○), 1.0×10^{-6} cm^2 s^{-1} (□), 1.0×10^{-7} cm^2 s^{-1} (△), 1.0×10^{-8} cm^2 s^{-1} (◇); other conditions as given in Figure 6C. (——) Theory according to the extended Golay equation (*33*).

The trends may be more clearly illustrated in graphical form, as shown in Figure 15. The rate constants increase linearly with the constant a up to approximately 0.5, whereafter the system becomes diffusion limited for these simulation conditions. From the rate constants given in Table IV, we can estimate the barrier to interfacial transport relative to the diffusion-limited case ($a = 1.0$). These barriers correspond to 0.17 $k_B T_0$ for $a = 0.5$, 1.70 $k_B T_0$ for $a = 0.1$, 2.36 $k_B T_0$ for $a = 0.05$, 3.92 $k_B T_0$ for $a = 0.01$, 4.61 $k_B T_0$ for $a = 0.005$, and 6.20 $k_B T_0$ for $a = 0.001$. From these calculations, it is evident that relatively small barriers can have a significant effect upon the kinetic behavior of the system. Consequently, we must seek to minimize sources of interfacial resistance to mass transport in order to develop unified chromatographic systems with optimal kinetic performance. This may involve minimizing surface tension effects, minimizing configurational or orientational effects, and choosing fluid phase solvents and modifiers that can be easily and rapidly disassociated from solute molecules at the interface.

The effect of interfacial resistance to mass transfer on the fluid dynamic behavior of the system has also been examined. The solute zone profiles obtained at a mean linear velocity of 0.10 cm s^{-1} are shown in Figure 16 for values of the constant a of 1.0, 0.1, 0.01, and 0.001. Although the characteristic time τ is very similar for these values of the constant a and for diffusion coefficients in the surface phase from 10^{-5} to 10^{-8} cm^2 s^{-1} shown in Table III, there is a marked difference in the solute zone profiles in Figures 12 and 16. The profiles for decreasing values of the constant a broaden but remain symmetric regardless of the magnitude of the characteristic time τ. Symmetry is preserved because the constant a influences both P_{fs} and P_{sf} in equation [10a,b] in the same manner. In contrast, the diffusion coefficients influence only one of the probability expressions in equation [10a,b] and mass transport in only one phase, resulting in zone profiles that are broader and more asymmetric as the characteristic time τ increases.

The statistical moments of the solute zone profiles are shown as a function of the simulation time in Figure 17. For the constant $a = 1.0$, the mean distance and variance agree well with the theoretically expected values. As the constant a decreases, the mean distance increases slightly because the molecules spend more time in the fluid phase before an effective transfer can occur at the interface. In general, the mean distance agrees reasonably well with the theoretically expected value for all except the smallest value of $a = 0.001$. However, the variance of the zone increases significantly as the constant a decreases. Only for the case of $a = 1.0$ does the variance conform to the extended Golay equation (*33*). This is because the Golay equation assumes that equilibrium conditions exist at the interface and does not consider the effects of slow interfacial mass transfer. These contributions rapidly become important as the constant a decreases and become predominant for values of a less than 0.01. It is also noteworthy that these contributions to variance increase linearly with simulation time, suggesting that steady-state conditions have been achieved. Clearly, interfacial resistance to mass transport in chromatographic systems is important and merits further study and more detailed characterization.

Conclusions

A three-dimensional stochastic simulation has been developed for the unified treatment of separation systems for chromatography, electrophoresis, and electrochromatography. This simulation follows the trajectories of individual molecules through the mass transport processes of diffusion, convection by laminar and electroosmotic flow, electrophoretic migration, and surface interaction by absorption and adsorption mechanisms. This simulation provides the opportunity to perform hypothetical experiments and to make observations that may not be possible with classical theoretical models or experiments.

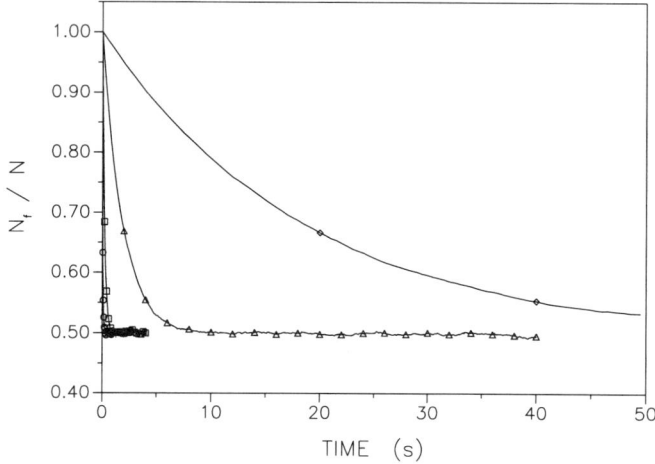

Figure 14. Kinetic evolution of the absorption process with varying interfacial resistance to mass transport. Simulation conditions: $D_f = 1.0 \times 10^{-5}$ cm^2 s^{-1}; $D_s = 1.0 \times 10^{-5}$ cm^2 s^{-1}; $a = 1.0$ (O), 0.1 (□), 0.01 (△), 0.001 (◇); other conditions as given in Figure 3B.

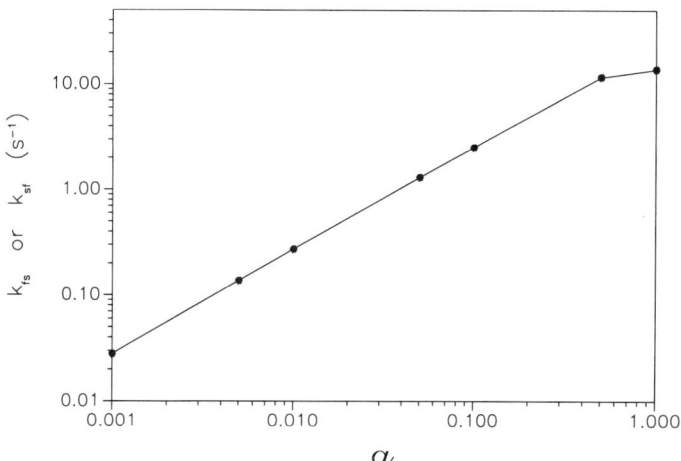

Figure 15. Effect of interfacial resistance to mass transport on rate constants k_{fs} (O) and k_{sf} (●). Simulation conditions as given in Table IV.

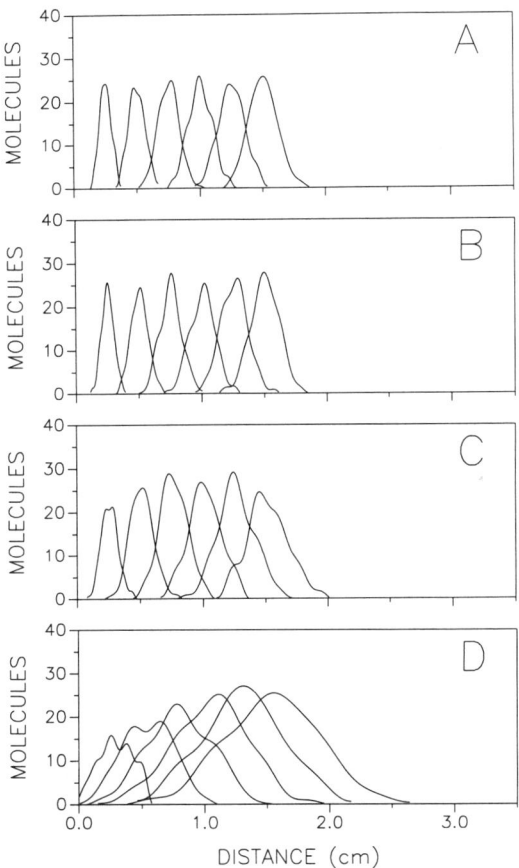

Figure 16. Evolution of the solute zone profile with varying interfacial resistance to mass transport for liquid chromatography. Simulation conditions: $D_f = 1.0 \times 10^{-5}$ cm^2 s^{-1}; $D_s = 1.0 \times 10^{-5}$ cm^2 s^{-1}; $a = 1.0$ (A), 0.1 (B), 0.01 (C), 0.001 (D); other conditions as given in Figure 6C.

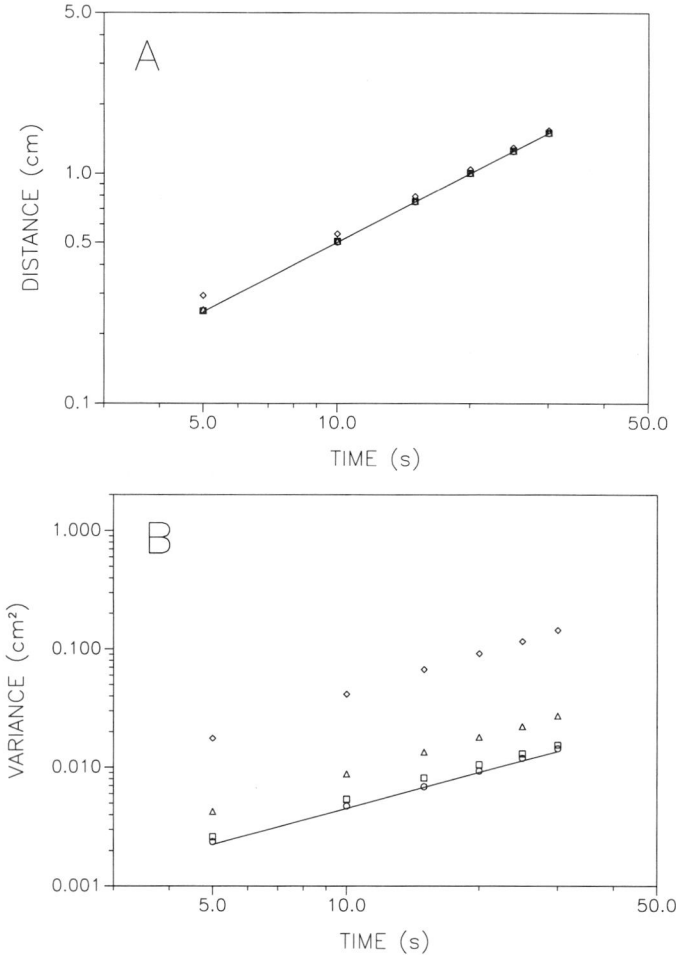

Figure 17. Mean distance (A) and variance (B) of the solute zone profile with varying interfacial resistance to mass transport. Simulation conditions: $D_f = 1.0\times10^{-5}$ cm^2 s^{-1}; $D_s = 1.0\times10^{-5}$ cm^2 s^{-1}; $a = 1.0$ (○), 0.1 (□), 0.01 (△), 0.001 (◇); other conditions as given in Figure 6C. (——) Theory according to the extended Golay equation (*33*).

In the present study, this simulation has been used to characterize the absorption mechanism between homogeneous fluid and surface phases for chromatography. The kinetic and equilibrium properties as well as the fluid dynamic behavior were examined as a function of the diffusion coefficient in the fluid and surface phases and the interfacial resistance to mass transport. These results suggest that an increase in surface phase diffusion coefficients and a reduction of interfacial resistance are the most critical factors in the development of more rapid separation systems.

Acknowledgments

This research was supported by the U.S. Department of Energy, Office of Basic Energy Sciences, Division of Chemical Sciences, under Contract No. DE-FG02-89ER14056. In addition, partial support for the IBM RS/6000 Model 580 computer was provided by the Michigan State University Office of Computing and Technology and the Center for Fundamental Materials Research.

Literature Cited

(1) Betteridge, D.; Marczewski, C. Z.; Wade, A. P. *Anal. Chim. Acta* **1984**, *165*, 227.
(2) Crowe, C. D.; Levin, H. W.; Betteridge, D.; Wade, A. P. *Anal. Chim. Acta* **1987**, *194*, 49.
(3) Wentzell, P. D.; Bowridge, M. R.; Taylor, E. L.; MacDonald, C. *Anal. Chim. Acta* **1993**, *278*, 293.
(4) Schure, M. R. *Anal. Chem.* **1988**, *60*, 1109.
(5) Schure, M. R.; Weeratunga, S. K. *Anal. Chem.* **1991**, *63*, 2614.
(6) Schure, M. R.; Lenhoff, A. M. *Anal. Chem.* **1993**, *65*, 3024.
(7) Hopkins, D. L.; McGuffin, V. L. *Anal. Chem.* **1998**, *70*, 1066.
(8) Guell, O. A.; Holcombe, J. A. *Anal. Chem.* **1990**, *62*, 529A.
(9) McGuffin, V. L.; Wu, P. *J. Chromatogr. A* **1996**, *722*, 3.
(10) McGuffin, V. L.; Krouskop, P. E.; Wu, P. *J. Chromatogr. A*, in press (1998).
(11) Giddings, J. C. *J. Chem. Ed.* **1958**, *35*, 588.
(12) Chen, J.; Weber, S. G. *Anal. Chem.* **1983**, *55*, 127.
(13) Press, W. H.; Flannery, B. P.; Teukolsky, S. A.; Vetterling, W. T. *Numerical Recipes: The Art of Scientific Computing*; Cambridge University Press: Cambridge, England, 1989.
(14) Einstein, A. *Ann. Phys.* **1905**, *17*, 549.
(15) Reid, R. C.; Prausnitz, J. M.; Sherwood, T. K. *The Properties of Gases and Liquids*; McGraw-Hill: New York, NY, 1977.
(16) Feller, W. *Probability Theory and its Application*; Wiley: New York, NY, 1950, Ch. 14.
(17) Krouskop, P. E. Unpublished research, Michigan State University, East Lansing, MI (1998).
(18) Taylor, G. *Proc. Roy. Soc. (London)* **1953**, *A219*, 186.
(19) Aris, R. *Proc. Roy. Soc. (London)* **1956**, *A235*, 67.
(20) Bird, R. B.; Stewart, W. E.; Lightfoot, E. N. *Transport Phenomena*; Wiley: New York, NY, 1960.
(21) Sherwood, T. K.; Pigford, R. L.; Wilke, C. R. *Mass Transfer*; McGraw-Hill: New York, NY, 1975.
(22) Hines, A. L.; Maddox, R. N. *Mass Transfer: Fundamentals and Applications*; Prentice-Hall: Englewood Cliffs, NJ, 1985.
(23) Karger, B. L.; Snyder, L. R.; Horvath, C. *An Introduction to Separation Science*; Wiley: New York, NY, 1973.
(24) Rice, C. L.; Whitehead, R. *J. Phys. Chem.* **1965**, *69*, 4017.

(25) Martin, M.; Guiochon, G. *Anal. Chem.* **1984**, *56*, 614.
(26) von Smoluchowski, M. *Bull. Intern. Acad. Sci. Cracovic* **1903**, *1903*, 184.
(27) Huckel, E. *Physik. Z.* **1924**, *25*, 204.
(28) McEldoon, J. P.; Datta, R. *Anal. Chem.* **1992**, *64*, 230.
(29) Atkins, P. W. *Physical Chemistry*; W. H. Freeman: San Francisco, CA, 1978.
(30) Robinson, R. A.; Stokes, R. H. *Electrolyte Solutions: The Measurement and Interpretation of Conductance, Chemical Potential, and Diffusion in Solutions of Simple Electrolytes*; Butterworths: London, England, 1959.
(31) Butler, J. N. *Ionic Equilibrium: A Mathematical Approach*; Addison-Wesley: Reading, MA, 1964.
(32) Laitinen, H. A.; Harris, W. E. *Chemical Analysis*, 2nd Ed.; McGraw-Hill: New York, NY, 1975.
(33) Golay, M. J. E. In *Gas Chromatography 1958*; Desty, D. H., Ed.; Academic Press: New York, NY, 1958, p. 36.
(34) Krouskop, P. E.; McGuffin, V. L. *J. Chromatogr.*, manuscript in preparation.
(35) Wu, P.; McGuffin, V. L. *AIChE J.* **1998**, *44*, 2053.

Chapter 5

Computer Simulations of Interphases and Solute Transfer in Liquid and Size Exclusion Chromatography

Thomas L. Beck and Steven J. Klatte

Department of Chemistry, University of Cincinnati, Cincinnati, OH 45221

This chapter summarizes molecular level modeling of chromatographic interphases. Previous studies are reviewed concerning chain structure and dynamics and solute retention, and new results are presented from computer simulations of liquid chromatographic interphases of C_{18} chains in contact with three different water/methanol mobile phases. These simulations probe the particle densities and free volume profiles across the interface, solvent orientation passing from bulk into the stationary phase, and dynamical properties of the alkane chains and solvent. Discussion is given of preliminary studies of the partitioning of charged solutes in size exclusion chromatography. Double layer effects are included by numerical solution of the nonlinear Poisson-Boltzmann equation which yields the potential of mean force between the charged dendrimer solute and the like-charged pore.

Chromatographic systems present many tough challenges for the researcher trying to obtain a fundamental understanding of the driving forces behind retention. Two prototype chromatographic systems are Reversed Phase Liquid Chromatography (RPLC) and Size Exclusion Chromatography (SEC) of charged particles. In each case the stationary phase is a complicated porous network and the surface is typically rough and has either been derivatized with long chain alkanes (RPLC) or is charged (SEC in a silica network). The mobile phase in RPLC most often is a complex fluid mixture of water/methanol or water/acetonitrile, while in SEC one may have a salt solution at a certain pH and ionic strength. In the SEC system, the ionic strength determines the effective size of charged solutes, which ultimately influences the partition coefficient.

Due to the relatively slow flow rates, one central theme in these separation strategies is that the retention can be modeled as an equilibrium process. For example, the retention factor in RPLC is simply the product of a phase ratio and an equilibrium constant for partitioning between the mobile and stationary phases. That partition coefficient in turn is directly related to the difference in solute excess chemical potentials between the two phases. This allows for some theoretical simplification, yet the factors which determine the equilibrium partition coefficient can be highly complex. A range of interactions such as electrostatic, hydrogen bonding, dispersion, and hydrophobic forces can all act together (or in competition) to yield the observed behavior.

In recent years, a great deal of effort has been focused on elucidating the underlying driving forces of retention in RPLC. A recent issue of the *Journal of Chromatography* (*1*) is devoted entirely to this topic. Experimentally, fundamental molecular level techniques such as fluorescence (*2,3*), NMR (*4,5*), and IR (*6*) spectroscopies have begun to unravel the structure and dynamics of the liquid chromatographic interface and the behavior of solutes there. Also, chromatographic separations themselves, through temperature and solvent dependences (*7,8*) and relations to bulk partitioning measurements (*9*), can suggest retention models and the relative importance of mobile and stationary phases. It is clear from the chromatographic research that the surface chain density influences retention (*10*), and the solvent composition also has a significant impact (*8*). Theoretically, several groups have developed molecular level approaches to rationalize the observed retention behavior. These theories have generally relied on mean field lattice statistical mechanics. The theories have led to a better understanding of the effect of surface chain density on retention (*11*), composition effects (*12*), and solute distributions into the stationary phase(*13*). However, the underlying physical assumptions in these models are restrictive due to the complexity of the aqueous mixtures and the tethered alkane stationary phases. For SEC separation of charged solutes, both experiment and analytical linearized Debye-Hückel level theory have recently been applied to controlled dendrimers (*14*). It was found that the linearized theory has significant shortcomings in explaining the observed effects of ionic strength on retention.

In this chapter, we discuss an alternative approach based on molecular level simulations (RPLC) and numerical Poisson-Boltzmann methods for SEC of charged solutes. While the molecular interactions employed in the simulations of RPLC interphases are not necessarily of sufficient accuracy to predict the very small excess chemical potential differences between similar solutes which can lead to separations, the models are accurate enough to lead to a detailed picture of the interfaces, namely their structure and dynamics, and they can yield semiquantitative estimates of excess chemical potential profiles for solute retention. The models are capable of including approximately all of the contributing interactions without recourse to mean field assumptions. In the SEC calculations, we can begin to assess the limitations of the linearization assumption (in Debye-Hückel theory) for prediction of retention and the importance of the ion valences in solution. The purposes of the research are to test the assumptions of the analytical models, to compare directly to experiment where possible, and hopefully to lead to a more accurate molecular level understanding of these interfaces. We do not review the extensive experimental literature nor analytical theoretical approaches, which are presented in other chapters of this volume and elsewhere. Rather we refer to the relevant experimental and theoretical results in relation to the molecular modeling calculations presented here.

Previous Simulations of RPLC Interphases

Our first simulations were of high density C_8 tethered alkane stationary phases in vacuum (15). The temperature dependent behavior (100-400K) was examined to explore the onset of any possible phase transitions. Disordered density profiles were observed at all temperatures, with some softening occurring with increasing temperature. The overall phase width was roughly 9Å, in agreement with neutron scattering data (16). The order parameter S_n was computed at all temperatures, and values near zero were observed, indicating a highly disordered system where the chains are on average tilted over towards the surface to a significant degree. No collective tilt was apparent, due to the large available free volume (the chains were bonded at roughly 1/2 closed packed density, typical of RPLC stationary phases). Even at low temperatures, a substantial fraction of the internal dihedral angles were in the gauche configuration (20%), in agreement with IR experiments (6). The chains undergo a gradual transition from a low temperature disordered, glassy state to a high temperature liquid-like form between 200K and 300K. Calorimetry experiments (17) indicate a broadened phase transition over this temperature range. The liquid-like state is characterized by diffusive motions of the chain segments, with increasing mobility further from the point of attachment. We computed diffusion constants over short time scales to differentiate between liquid and solid or glassy behavior. We note here that any value between our computed diffusion constants and zero can be obtained for tethered chains, depending on the time scale of the calculation for the slope of the mean square displacement, since the chains are not free to roam on the surface (18).

Our second series of simulations examined chain length, surface density, and intermolecular force effects on stationary phases (19). The density profiles for longer chain systems exhibit a clear layering away from the surface, even in a liquid-like state. This layering has also been observed for free long chain alkanes near solid surfaces (20). The segment density tends to fill the available extensive free volume, which amounts to at least 50% even at the highest available commercial stationary phase packings. This is not surprising since octadecane on a surface is expected to be liquid or glassy at these temperatures. This is the reason the chains do not assume completely extended conformations on average, and the interphase width is substantially less than the fully extended value. The role of each attractive component of the chain and surface potentials was explored by sequentially simulating each of the four combinations of purely repulsive and full Lennard-Jones interactions. Both surface and intersegment attractive potentials are crucial in obtaining a realistic model of the interfacial structure. This result indicates that purely repulsive theoretical chain models fail to capture important features of the interactions. Diffusion constants were computed as a function of distance from the solid surface, and for C_{18} stationary phases, the first 10Å of the stationary phase is largely 'frozen' into a glassy state on the surface. The upper 8-10Å region exhibits liquid-like behavior at room temperature. Variation of the state of the stationary phase with distance from the surface may have a large impact on retention behavior. Topographic maps were displayed of the C_{18} stationary phase surfaces at several densities and over a time sequence at a typical RPLC density. While the chain tails exhibit short time liquid-like motions, large scale features on the surface move on a much slower time scale (on the order of 100ps). The surface is rough and disordered, which has been inferred from spectroscopic experiments on a large hydrophobic solute (*3*).

Water/methanol solvent mixtures were then added to the modeled stationary phases (*21*). The system configurations consisted of two opposing stationary phase layers with the solvent mixture in between. The overall interphase width was 100Å in order to obtain bulk behavior in the center of the simulation cell. A total of roughly 8000 force centers were required to obtain a system of this size. Periodic boundary conditions were enforced in two dimensions. A small hydrophobic solute (methane) was then inserted into a range of slabs at distances ranging from bulk solution into the stationary phase. The system examined corresponds to a 50:50 water/methanol mixture (by volume) and a C_{18} stationary phase at 4 μmol/m². A window potential technique was used to compute the potential of mean force (excess chemical potential profile) for motion of the solute into the stationary phase. The profile displays a small peak at the interface (of magnitude roughly kT, and presumably due to ordered methanol, see below), and then there is a decrease as the solute passes into the stationary phase. The bulk to minimum difference is -3.6 kcal/mol. There are two minima in the stationary phase region (solute completely surrounded by chain segments) due to the layering effect discussed previously, and then the free energy rises sharply due to the frozen and high density alkane layers near the surface. If a local average of the free energy were taken around the minimum in the stationary phase (a solute 'sees' a range of free energies in the interfacial zone), then an overall free energy change of around -3.0 kcal/mol would be expected from these calculations. This value is relatively close to what would be expected for bulk partitioning of methane between water/methanol and oil (roughly -2 to -2.5 kcal/mol) (*22,23*). A somewhat related value is that for retention of benzene (-3.05 kcal/mol) for a 95:5 water/propanol mobile phase and 2.39 μmol/m² stationary phase (*7*), which is quite close to the bulk water/alkane partitioning value of -3 kcal/mol (*9*). Recently, preliminary measurements have been made (Wysocki, J.; Dorsey, J. G.; Beck, T. L., in progress) for retention of methane on a 50:50 water/methanol mobile phase, 4.1 μmol/m² stationary phase column which display evidence of the hydrophobic effect on solubilities (in the temperature dependence). The initial estimate of the overall free energy drop is -0.8 kcal/mol, but as of this writing, this value can only be considered a preliminary estimate; it is lower in magnitude than what would be expected for a bulk partitioning process.

The simulation results show that a partitioning mechanism occurs, but the retention cannot be considered a truly bulk partitioning process since the free energy profile never levels in the stationary phase region. There are specific ordering and dynamical effects there that prevent the solute from seeing a bulk alkane fluid environment. However, this does not preclude the possibility that the overall free energy changes for nonpolar solutes are very close to those expected for bulk partitioning. In fact, the recent experiments of Carr, et al. (9) show there is a close correlation between RPLC retention free energies and bulk partitioning for nonpolar solutes (for long chain and/or high density stationary phases and water/methanol mobile phase compositions which are below 70% methanol).

In related simulation studies, Schure modeled stationary phases in vacuum and in contact with water/methanol mobile phases (24). He created a nonuniform surface phase with higher chain concentrations in the middle of the cell. Chain density profiles are not presented, but single snapshots are shown of chain conformations. He observes highly extended chain conformations in contradiction to our findings for RPLC densities. No order parameters were presented to give a quantitative estimate of the average degree of chain tilting relative to surface normal. In simulations with solvent, preferential segregation of methanol into the stationary phase was observed for the 80% methanol case, which agrees with our findings (below). Finally, he presents radial distribution functions for decane interacting with water/methanol mixtures, and sees a significant depletion of water from around this large hydrophobic solute. This effect was not observed in our simulations of methane in bulk water/methanol, which is likely due to the much smaller size of methane.

Yarovsky, et al. (25) presented extensive simulations of alkane stationary phases in vacuum. The layer thickness was calculated for three bonding densities for C_4, C_8, and C_{18} systems. However, this quantity was calculated as the average distance of the terminal methyl group to the surface. Previous simulations (19) have shown that the terminal methyl exhibits a wide range of motions perpendicular to the surface, so it is likely not an accurate indicator of overall phase width; the total chain segment density profile is a better indicator of the full alkane width for comparison with neutron scattering experiments. They also computed the average number of gauche defects for the different chain lengths and surface densities. For the highest density examined (3.69 μmol/m^2), they find that extended chain conformations predominate for C_{18} stationary phases based on end-to-end distributions. Diffusion constants were computed which illustrate the increased mobility for segments away from the surface. Values significantly lower than ours were obtained since they computed the diffusion constant over much longer time intervals. Recently, Yarovsky, et al. (26) have extended their studies to larger systems of peptides in contact with alkane stationary phases.

Martin and Siepmann (27) have utilized a combination of Configurational Bias (CBMC) and Gibbs Ensemble (GEMC) Monte Carlo techniques to study multicomponent phase equilibria and solute partitioning. They employed three different alkane force fields to assess the accuracy of each. The authors examined the boiling point diagram for an n-octane n-dodecane mixture, an isotherm for a binary supercritical mixture of ethane and n-heptane, and free energies for solute (n-pentane, n-hexane, and n-heptane) partitioning between a helium vapor phase and an n-heptane liquid phase. For the solute transfer studies, the computed free energies are within 0.5 kJ/mol of the experimental values. These simulations show that complex fluid mixtures can be realistically modeled with current intermolecular potentials and simulation methods. The quantitative agreement with experiment is impressive; extensions to aqueous mixtures provide more severe challenges to molecular models due to the strong electrostatic forces.

Computational Methods

Our molecular dynamics simulation methodology has been described in detail elsewhere (*15,19,21*). Simulations of the tethered alkane water/methanol interphases were performed in the constant energy, constant volume ensemble. The chain segment (and solute or solvent) surface interactions were modeled as an integrated Lennard-Jones interaction to include the average long range attractions (*28*). The well depth parameters were estimated from heat of adsorption data of alkanes on silica. Two surfaces were modeled: one which included roughness on a 4Å scale via an oscillatory potential and one flat. In the figures below the rough surface is on the left. The united atom alkane interactions were taken from simulations of related monolayers (*28*). The SPC potential (*29*) was utilized for water, while the OPLS parameters (*30*) were assumed for methanol (and the methane solute discussed above). The electrostatic interactions were smoothly truncated group by group starting at 8.5Å with a switching function. The alkane parameters were truncated at 9Å. A total of roughly 8000 force centers were necessary to model an interphase width of 100Å in order to ensure bulk solvent behavior in the middle. There were 32 chains on each side (4 μmol/m^2). The systems were initially equilibrated with multiple heating/cooling annealing cycles before final equilibration and production runs at room temperature. Typical production runs were over at least 100 ps time scales.

In the numerical studies of charged solute partitioning into like-charged pores, the nonlinear Poisson-Boltzmann equation was solved:

$$\nabla \cdot (\epsilon(\mathbf{r})\nabla\phi(\mathbf{r})) = -4\pi[\rho_s(\mathbf{r}) + q\bar{n}_+ e^{-\beta q\phi(\mathbf{r})-v(\mathbf{r})} - q\bar{n}_- e^{\beta q\phi(\mathbf{r})-v(\mathbf{r})}] \quad (1)$$

where $\phi(\mathbf{r})$ is the electrostatic potential, $\beta = 1/kT$, $\epsilon(\mathbf{r})$ is the (possibly) spatially dependent dielectric constant, $\rho_s(\mathbf{r})$ is the discrete source charge density, q is the charge on the solution ions, $\bar{n}_\pm$ are the concentrations of the positive and negative ions in zero potential regions, and $v(\mathbf{r})$ is a very large, positive excluded volume potential. This equation is a mean field (or saddle point) solution to an exact functional integral expression for the partition function of the ion gas. We solved it numerically with a highly efficient multigrid method (*31*; Coalson, R. D.; Beck, T. L. In *Encyclopedia of Computational Chemistry*; in press). Once the electrostatic potential is obtained over the whole domain, the Helmholtz free energy of the ion gas can be computed. By computing this free energy for several solute configurations, the potential of mean force profile can be obtained, and from this the partition coefficient:

$$K = 2 \int_0^{1-\delta} \exp[-\beta A(\alpha)] \alpha \, d\alpha \qquad (2)$$

where α is the distance from the center scaled by the pore radius, $A(\alpha)$ is the free energy at that distance, and δ is the sphere radius divided by the pore radius. Note, for a noninteracting particle of zero radius, the partition coefficient is one.

In our calculations, we modeled a charged sphere (dendrimer) passing from bulk solution into a like-charged cylindrical pore, so periodic boundaries must be employed. Some subtlety is involved in computing the free energy for the periodic case, since a recently proposed variational form (*32*) is not invariant to a uniform shift in the electrostatic potential. We have derived the connection between the field theory free energy and the variational result which illustrates the connections and differences (Beck, T. L.; Coalson, R. D., in preparation). The computation is a two step process: first compute the free energy to move the sphere to the center of the cylinder, then compute the free energy profile to move it out towards the cylinder. A uniform dielectric was assumed in our calculations (monovalent salts at 0.03M ionic strength), so no image forces were included (this should not lead to significant errors under these conditions). The dielectric constant was taken as that of water, and a grid spacing of 7.19Å was used. The total size of the cubical box on one side was 460Å. Additional inputs to the program are the sphere and cylinder surface charge densities and the total numbers of positive and negative ions.

Results and Discussion

Results for molecular dynamics simulations of RPLC interphases are presented first. The relative particle density profiles across the full interphase are shown in Figure 1. The chain density profiles (the average along z of all alkane united atom segment densities) are clearly visible on either side of each Figure. The left side is the rough surface, while the right side is flat. The surface undulations on the left side disrupt to some extent the horizontal layering of the chains visible in Figures 1a,c (highly aqueous mobile phases). This is expected since these density profiles are for cuts along the z direction, and not in terms of local distance from the hard wall; even if there were strong local density oscillations away from the surface, these would be smoothed by the averaging process employed here. The same argument pertains to the solvent density profiles which respond to local density oscillations of the stationary phase, so we prefer to emphasize the density profiles near the flat surface on the right. (The chains were bonded via a random growth process and were assigned random z points of attachment within a 2Å window to mimic roughness even on the flat solid surface). In the central region, the water density (density of Lennard-Jones particles centered on the oxygen) is the top curve in Figures 1a and 1c, and the methanol density (density of Lennard-Jones particles centered on the methyl group) is the top curve in Figure 1b. The bulk water/methanol number densities correspond closely to the experimental values for these three mixtures.

For the two cases in Figures 1a and 1c, the alkane density profile near the flat

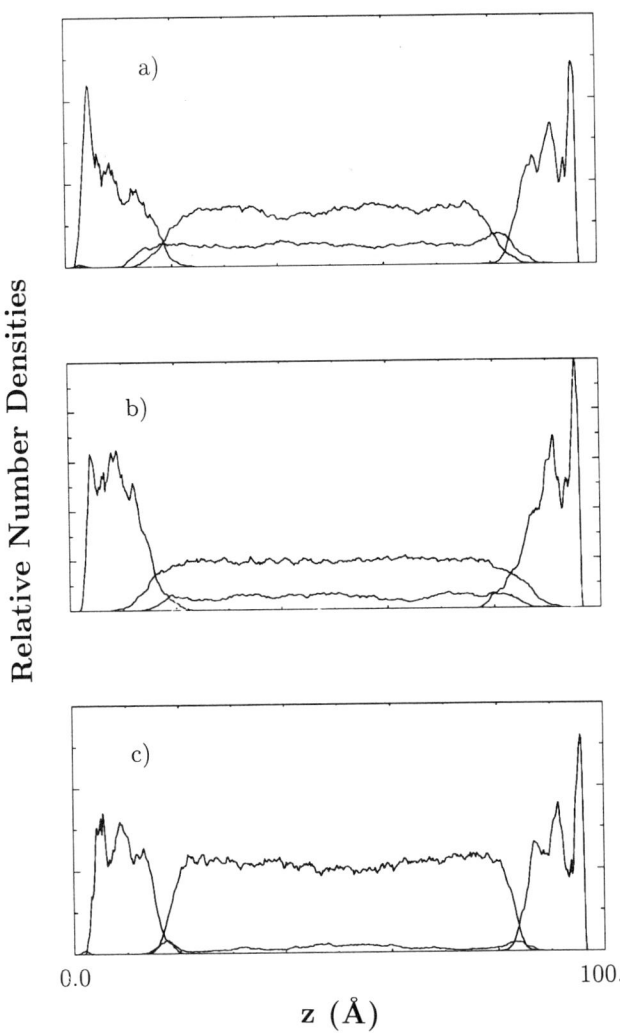

Figure 1. Relative number densities for chain segments, water, and methanol. a) 50:50 (volume) mobile phase. b) 10:90 water/methanol. c) 88:12 water/methanol. The oscillatory C_{18} alkane chain segment densities are clearly visible on either side of each of the figures. For the solvent profiles, water is the top curve at the center (z=50Å) in a), methanol is the top curve in b), and water is the top curve in c).

surface is very close to that observed for the corresponding 4 μmol/m^2 stationary phase in vacuum (*19*). In other words, the aqueous, hydrogen bonded fluid does not seriously perturb the overall structure of the stationary phase. The chains exist in a relatively collapsed state on the surface, and cover the available silica surface with high density layers. Near the surface the chain segments are 'frozen' into a glassy state, so this layer is largely impenetrable to solvent and presumably solute. The chain tails exhibit liquid-like motions; however, the computed short time diffusion constants are suppressed up to 50% relative to the chains in vacuum. The outer surface of the stationary phase is rough (and fluxional) on a 10Å length scale. For the 50:50 and 88:12 water/methanol mixtures it is apparent, especially on the side of the flat surface, that methanol segregates strongly to the surface. In effect, methanol acts like a 'mini-surfactant', which is consistent with the fact that it reduces the surface tension relative to pure aqueous systems. For the high methanol content case (Figure 1b), the methanol penetrates substantially further into the stationary phase, and the outer layer of the chain density profile disappears as the chain density protrudes into the solvent. The water density does not penetrate significantly into the stationary phase for any of the three concentrations (There is some overlap of the density profiles due to the undulations of the fluid surfaces. This effect has been observed in simulations of water/alkane interfaces, where the local interfacial width is quite small (*33*)). Computed solvent diffusion constant profiles show some suppression near the alkane interfaces, on the order of 10-20%, with more reduction in the z direction than in the xy plane.

The effect of chain solvation is apparent in Figure 2, which displays the occupied volume along the z profiles corresponding to Figure 1. This volume was computed simply by binning the volumes of all Lennard-Jones centers in windows along the z direction. Dips in the occupied volume give a crude indication of free volume increases. There are clear indications of the repulsive nature of the interfaces for the 50:50 and 88:12 water/methanol cases. Both are in the concentration range referred to as 'highly aqueous' in the RPLC literature. Simulations of water/alkane surfaces exhibit the same free volume increase at the interface (*23*). For the high methanol content phase the free volume excess disappears in the interfacial region, indicating the increased mixing of solvent and stationary phase. Octadecane is not soluble in methanol, but there is an increased interfacial width due to a partial solubilization at the interface.

An interesting issue is the relative orientation of the solvent passing from bulk towards the stationary phase. No special orientation of water was observed. However, the methanol becomes highly orientationally ordered near the interface for the 50:50 mixture (Figure 3). In this figure the distributions of angles of the methanol O-C axis relative to the surface normal are plotted for several z windows. These distributions are computed to give a flat curve in the bulk, but are not normalized for the number of particles. Therefore, curves with smaller magnitudes indicate fewer methanols in that slab. It is evident from this Figure that methanols near the alkane interface are ordered with the methyl group pointing toward the hyrdrophobic surface. This orientational behavior is in agreement with

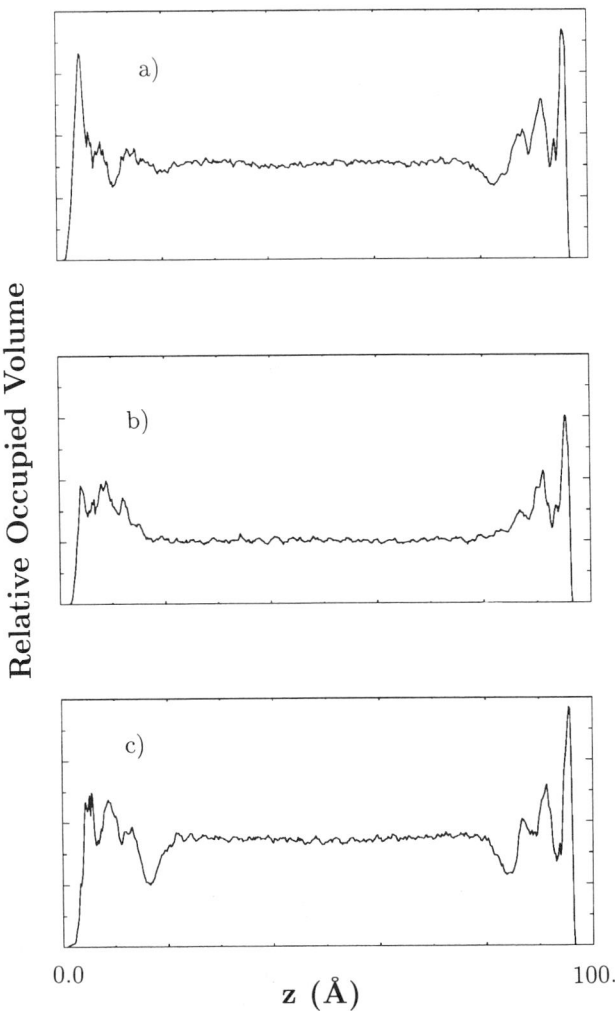

Figure 2. Total occupied volume along the z profile. a) 50:50 (volume) mobile phase. b) 10:90 water/methanol. c) 88:12 water/methanol. The large oscillations at either side are due to the alkane chains, while the free volume increase (or occupied volume decrease) is apparent at the solvent/stationary phase interface, for example near z=85Å in c).

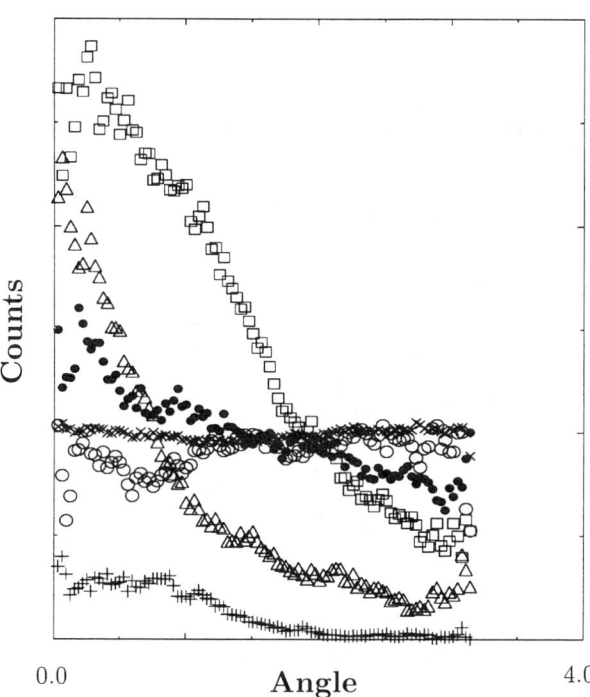

Figure 3. Distribution of methanol orientational angles relative to surface normal in windows centered at several z locations (see Figure 1a). The distribution range is 0 to π. Bulk (×); 70 Å (open circles); 77 Å (filled circles); 82 Å (open squares); 86 Å (open triangles); 88 Å (+). The relatively flat and low magnitude curve for the last window is due to a few trapped methanols not in contact with the bulk solvent.

sum-frequency vibrational spectroscopy measurements on free water/methanol liquid-vapor interfaces (*34*).

The simulation results are consistent with a growing body of experimental results on RPLC water/methanol interphases. Harris and coworkers (*2*) observed collapse of the stationary phase in highly aqueous mixtures. Montgomery, et al. (*3*) found that a large hydrophobic solute resides at the interface between mobile and stationary phases, and that the interface is rough. Sander, et al. (*6*) performed IR measurements of stationary phase chains in contact with solvent, and found that for high methanol content, there is a *partial* extension of the chains, consistent with our result. Bliesner and Sentell (*5*) monitored motions of water and methanol (also water/acetonitrile) via deuterium NMR measurements. They observed a strong association of the methanol with the stationary phase, especially for high methanol fractions. They did not find evidence of solvent penetration and binding to the silica surface, at least for relatively high chain densities. In addition, Carr and coworkers (*9*) have recently performed a series of elegant chromatographic experiments which show that *free energies* for retention of nonpolar solutes can be accurately modeled as a bulk partitioning process. Deviations from agreement occurred for high methanol fractions, and they postulated this may be due to penetration of methanol into the stationary phase environment, thus altering the local excess chemical potential of the solute there. Our simulations support that interpretation.

Finally, we present preliminary results pertaining to SEC of charged dendrimers. The system parameters were chosen to conform to the G3 dendrimers at 0.03M ionic strength in the experimental system (*14*). The observed experimental partition coefficient is $K = 0.30$. (The extreme value of completely screened hard sphere solute is $K = 0.63$). We modeled only monovalent ions in our initial calculations. In Figure 4, the charge density contours of salt negative ions in the pore are plotted. The solute is partially along the path from the center of the pore to the side. The overlap of the sphere and cylinder double layers is apparent in the figure. The net free energy for this configuration (relative to separate sphere and cylinder) is $0.56kT$ at room temperature. The free energy was computed for each configuration, and from the assembled values a partition coefficient of $K = 0.123$ was obtained. This value is slightly more accurate than the Debye-Hückel result obtained by Shah, et al. (*14*) ($K = 0.11$) but is still far below the experimental value. Thus inclusion of nonlinear effects appears to improve the result but only slightly. One important factor to note is that both monovalent and divalent ions were used in the buffer solution. The deviation of the theoretical result could be due to several factors: inaccurate input values for surface charge densities, the assumption of spherical and cylindrical geometries for solute and pore, lack of variable dielectric constant, charge regulation at the surfaces, failure of the Poisson-Boltzmann level of theory, and/or use of only monovalent ions. The last factor is most likely the major source of the error, since divalent ions can lead to large reductions in repulsive interactions (or even attractions) between like charged surfaces (*35*). Future studies will focus on these issues, and simulations of flexible polyelectrolyte chains in pores using a new Multigrid Configurational

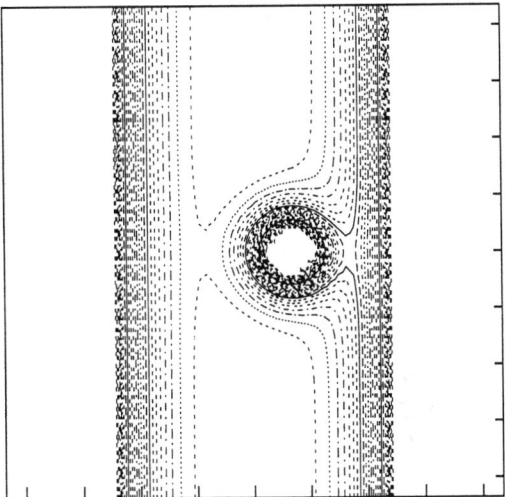

Figure 4. Contour plot of the negative salt ion charge density for a negatively charged dendrimer (sphere) inside a cylindrical silica pore. The total width is 460 Å in each direction. The excluded volume solid region (no mobile ion density) is visible on either side of the cylindrical pore.

Bias Monte Carlo method for those calculations (Liu, A.; Beck, T. L.; Frenkel, D., in preparation).

Conclusions

Computer simulations of RPLC interphases and solute transfer have led to more detailed pictures of the molecular level events at the interface between the mobile and stationary phases. Generally good agreement is obtained between experimental findings and the calculations. The research shows that the current level of molecular models and computational methodolgy and resources can be used to model the interfaces on a relatively accurate level. Much more work needs to be done, including: improvement of interaction potentials, simulations of free energies, enthalpies, and entropies for a wide range of nonpolar and polar solutes, and extensions to larger systems. Through combinations of theory, modeling, and experiment, we are beginning to gain a quantitative understanding of the driving forces and mechanisms of RPLC retention. The numerical calculations of electrostatic effects in SEC are just beginning, but hold promise to aid in understanding the important parameters for retention of charged molecules and polymers in porous media. New Monte Carlo methods will allow the computation of excess chemical potentials of charged flexible chains inside like-charged pores.

Acknowledgments. This research has been supported by the National Science Foundation, the donors of the Petroleum Research Fund of the ACS, and the Ohio Supercomputer Center.

Literature Cited

1. *J. Chromatogr.* **1993**, *656*.
2. Wong, A. L.; Hunnicutt, M. L.; Harris, J. M. *Anal. Chem.* **1991**, *63*, 1076.
3. Montgomery, M. E.; Green, M. A.; Wirth, M. J. *Anal. Chem.* **1992**, *64*, 1170.
4. Ziegler, R. C.; Maciel, G. E. *J. Phys. Chem.* **1991**, *95*, 7345; *J. Am. Chem. Soc.* **1991**, *113*, 6349.
5. Bliesner, D. M.; Sentell, K. B. *Anal. Chem.* **1993**, *65*, 1819.
6. Sander, L. C.; Callis, J. B.; Field, L. R. *Anal. Chem.* **1983**, *55*, 1068.
7. Cole, L. A.; Dorsey, J. G.; Dill, K. A. *Anal. Chem.* **1992**, *64*, 1324.
8. Alvarez-Zepeda, A.; Barman, B. N.; Martire, D. E. *Anal. Chem.* **1992**, *64*, 1978.
9. Tan, L. C.; Carr, P. W. *J. Chromatogr. A* **1997**, *775*, 1; Carr, P. W.; Li, J.; Dallas, A. J.; Eikens, D. I.; Tan, L. C. *J. Chromatogr. A* **1993**, *656*, 113.
10. Sentell, K. B.; Dorsey, J. G. *Anal. Chem.* **1989**, *61*, 930.
11. Dorsey, J. G.; Dill, K. A. *Chem. Rev.* **1989**, *89*, 331.
12. Martire, D. E.; Boehm, R. E. *J. Phys. Chem.* **1983**, *87*, 1045; Boehm, R. E.; Martire, D. E. *ibid.* **1994**, *98*, 1317.
13. Bohmer, M. R.; Koopal, L. K.; Tijsenn, R. J. *J. Phys. Chem.* **1991**, *95*, 6285.
14. Shah, G.; Dubin, P. L.; Kaplan, J. I.; Newkome, G. R.; Moorefield, C. N.; Baker, G. R. *J. Coll. Inter. Sci.* **1996**, *183*, 397.

15. Klatte, S. J.; Beck, T. L. *J. Phys. Chem.* **1993**, *97*, 5727.
16. Sander, L. C.; Glinka, C. J.; Wise, S. A. *Anal. Chem.* **1990**, *62*, 1099.
17. Van Miltenburg, J. C.; Hammers, W. E. *J. Chromatogr.* **1983**, *268*, 147.
18. Beck, T. L. *Anal. Chem.* **1996**, *68*, 1973.
19. Klatte, S. J.; Beck, T. L. *J. Phys. Chem.* **1995**, *99*, 16024.
20. Balasubramanian, S.; Klein, M. L.; Siepmann, J. I. *J. Phys. Chem.* **1996**, *100*, 11960.
21. Klatte, S. J.; Beck, T. L. *J. Phys. Chem.* **1996**, *100*, 5931. For a complete discussion of the methods and results of the simulation work, see the thesis of S. J. Klatte located at http://bessie.che.uc.edu/tlb/beck.html.
22. Ben-Naim, A. *J. Phys. Chem.* **1967**, *71*, 4002.
23. Pohorille, A.; Wilson, M. A. *J. Chem. Phys.* **1996**, *104*, 3760.
24. Schure, M. R. In *Chemically Modified Surfaces*; Pesek, J. J. and Leigh, I. E., Eds.; Society of Chemistry Information Services: Cambridge, UK, 1994.
25. Yarovsky, I.; Aquilar, M. I.; Hearn, M. T. W. *Anal. Chem.* **1995**, *67*, 2145.
26. Yarovsky, I.; Hearn, M. T. W.; Aquilar, M. I. *J. Phys. Chem. B* **1997**, *101*, 10962.
27. Martin, M. G.; Siepmann, J. I. *J. Am. Chem. Soc.* **1997**, *119*, 8921.
28. Bareman, J. P.; Klein, M. L. *J. Phys. Chem.* **1990**, *94*, 5202.
29. Berendsen, H. J. C.; Postma, J. P. M.; van Gunsteren, W. F.; Hermans, J. In *Intermolecular Forces*; Pullman, B., Ed.; Reidel Publishing Co.: New York, NY, 1981; p. 331.
30. Jorgensen, W. L.; Madura, J. D.; Swenson, C. J. *J. Am. Chem. Soc.* **1984**, *106*, 6638.
31. Beck, T. L. *Intl. J. Quant. Chem.* **1997**, *65*, 477.
32. Sharp, K.; Honig, B. *J. Phys. Chem.* **1990**, *94*, 7684.
33. Pohorille, A.; Wilson, M. A. *J. Mol. Struct.* **1993**, *284*, 271.
34. Wolfrum, K.; Graener, H.; Laubereau, A. *Chem. Phys. Letts.* **1993**, *213*, 41.
35. Guldbrand, L.; Jönsson, B.; Wennerström, H.; Linse, P. *J. Chem. Phys.* **1984**, *80*, 2221.

Chapter 6

Exploring Multicomponent Phase Equilibria by Monte Carlo Simulations: Toward a Description of Gas–Liquid Chromatography

Marcus G. Martin[1], J. Ilja Siepmann[1,3], and Mark R. Schure[2]

[1]Department of Chemistry, University of Minnesota,
207 Pleasant Street SE, Minneapolis, MN 55455-0431
[2]Theoretical Separation Science Laboratory, Rohm and Haas Company,
727 Norristown Road, Spring House, PA 19477

The calculation of retention times, retention indices, and partition constants is a long sought-after goal for theoretical studies in gas chromatography. Although advances in computational chemistry have improved our understanding of molecular interactions, little attention has been focused on chromatography, let alone calculations of retention properties. Configurational-bias Monte Carlo simulations in the Gibbs ensemble have been used to calculate single and multi-component phase diagrams for a variety of hydrocarbon systems. Transferable force fields for linear and branched alkanes have been derived from these simulations. Using calculations for helium/n-heptane/n-pentane systems, it is demonstrated that this approach yields very precise partition constants and free energies of transfer. Thereafter, the partitioning of linear and branched alkane solutes (with five to eight carbon atoms) between a squalane liquid phase and a helium vapor phase is investigated. The Kovats retention indices of the solutes are calculated directly from the partition constants.

The underlying principles of chromatographic separation are inherently complex, being dictated by the interplay of the sample with the stationary phase (solid substrate and bonded phase) and the mobile phase that often contains a mixture of solvents. Thus predicting the retention characteristics of a solute molecule given only its structure and the experimental chromatographic conditions is one of the grand challenges in separation science. Many different methods for the prediction of retention data have appeared in the literature (for excellent reviews, see (1-5)). We note here that many attempts at this predictive capability have

---

[3]Corresponding author: siepmann@chem.umn.edu.

inevitably ended up as some form of chemometric exercise using techniques like factor analysis or pattern recognition, i.e. equations are derived from fitting to experimental data (sometimes even retention data) for a series of molecules (the training set) and then used to interpolate or extrapolate for compounds not included in the series. A variety of equation-of-state based methods (6) and group additivity methods (7) have also been used for this purpose, and although based on composition, they do not explicitly consider the details of the molecular structure.

In this review article, we describe our progress using a more direct approach, that is we perform molecular simulations to explore the fluid phase equilibria that govern retention in gas-liquid chromatography (GLC). The simulated systems consist of stationary and mobile phases plus solute molecules. Use of novel simulation techniques and transferable force fields allows us to directly determine solute partitioning coefficients (or constants) and to predict relative retention times, and no information on other (macroscopic) physical data of the constituents, such as vapor pressure, refractive index or hydrogen acceptor/donor scale, is required.

The remainder of this article is divided as follows. The next section is devoted to a brief description of the simulation methodology. In the following section, simulation results for single-component alkane systems are described that were used to fit the force field parameters. Thereafter, the ability of the simulation methodology to yield precise vapor-to-liquid-phase free energies of transfer is demonstrated for mixtures of medium-length alkanes. Finally, initial results for the partitioning of linear and branched alkane solutes in a helium/squalane GLC system are presented.

Simulation Methodology

The position of chemical equilibria as well as the direction of all spontaneous chemical change is determined by free energies. The partition constant of solute S between phases α and β is directly related to the Gibbs free energy of transfer (8)

$$\Delta G_S = RT \ln \left(\frac{\rho_S^\alpha}{\rho_S^\beta} \right) \quad (1)$$

where ρ_S^α and ρ_S^β are the number densities of S in the two phases at equilibrium, and the ratio of these is the partition constant K. Whereas the determination of mechanical properties is now routine for computer simulation, the determination of (relative and absolute) free energies and other thermal properties, which depend on the volume of phase space, remains one of the most challenging problems (9,10). Many excellent reviews devoted to free energy calculation methods have appeared over the last ten years (e.g., see (11-13)).

Thermodynamic integration (TI) and free energy perturbation (FEP) are the most widely used methods to calculate free energy differences and are available in some commercial simulation packages. The TI method is based on the statistical connection between the Gibbs free energy, G, and the Hamiltonian, H, of the system

$$\Delta G = G_B - G_A = \int_0^1 \frac{\partial G_\lambda}{\partial \lambda} d\lambda = \int_0^1 \left\langle \frac{\partial H_\lambda}{\partial \lambda} \right\rangle_\lambda d\lambda \qquad (2)$$

where λ is a coupling parameter so that $H_{\lambda=0} = H_A$ and $H_{\lambda=1} = H_B$. The angular brackets, $\langle \cdots \rangle_\lambda$, denote an isobaric-isothermal ensemble average on the state defined by λ. Use of the TI expressions requires a reversible thermodynamic path. In the FEP method the difference between two state is treated as a perturbation (requiring that the states are not too different)

$$\Delta G = G_B - G_A = -RT \ln \left\langle \exp[-(H_B - H_A)/RT] \right\rangle_{P,T} \qquad (3)$$

To alleviate the problem that most often the initial and final states are quite different, A and B can also be linked through a number of intermediate states. Recently, Kollman and co-workers (14,15) investigated the precision of TI and FEP calculations and concluded that molecular dynamics simulations on the order of nanoseconds are required to give free energies of solvation of small molecules (methane) with a relative error of around 10%, and even longer simulations are needed to obtain reliable potentials of mean force.

None of the traditional free energy methods described in the preceding section is suitable for the efficient calculation of phase diagrams and until recently computation of multicomponent phase equilibria involving complex molecules was considered beyond reach (9). However, over the past few years many new methods have been proposed which greatly aid in the calculation of phase equilibria (16-22). In the interest of brevity, we describe only the salient features of the simulation methods which will be used in this work. The work-horse for our calculations is the combination of the Gibbs-ensemble Monte Carlo (GEMC) method (23-25) and the configurational-bias Monte Carlo (CBMC) algorithm (26-30). GEMC utilizes two separate simulation boxes that are in thermodynamic contact, but do not have an explicit interface. As a result, for a given state point the properties of the coexisting phases, such as the partitioning of solute molecules, can be determined directly from a single simulation. Since GEMC samples directly the partitioning between two phases, it can also be used to calculate the partition constant from the ratio of the number densities or, if so desired, also using the molality scale. Knowing the partition constant, K, the free energy of transfer, $\Delta G°$, can be calculated directly from eqn. 1, which is the same procedure as used for experimental data.

One of the GEMC steps involves the swapping of a molecule from one phase to the other, thereby equalizing the chemical potentials of each species in the two phases. Acceptance of these particle interchanges is often the rate limiting step in GEMC simulations and to improve the sampling of insertions of flexible molecules, such as the alkanes, the CBMC technique is used. CBMC replaces the conventional random insertion of entire molecules with a scheme in which the chain molecule is inserted atom by atom such that conformations with favorable energies are preferentially found. The resulting bias in the CBMC swap step, which improves the efficiency of the simulations by many orders of magnitude, is then removed by using special acceptance rules (*31,32*).

The extension to multicomponent mixtures introduces a case where the smaller molecules (members of a homologous series) have a considerably higher acceptance rate in the swap move than larger molecules. In recent simulations for alkane mixtures (*33*), we took advantage of this by introducing the CBMC-switch move which combines the GEMC identity switch (*24*) with CBMC chain interconversion (*34*), such that molecule A is regrown as molecule B in one box, and B is regrown as A in the other box. This CBMC-switch equalizes the difference of the chemical potentials of A and B in the two boxes. The conventional CBMC swap is used mainly to equalize the chemical potential of the shorter alkane in the two phases. This scheme is extremely beneficial for simulations of chromatographic systems, where many members of a homologous series are considered as solutes.

Single-Component Phase Equilibria and Force Field Development

For the past five years, our group has employed configurational-bias Monte Carlo simulations in the Gibbs ensemble to derive the force field parameters for non-bonded interactions (*30,35-38*). The united-atom version (in which entire methyl and methylene units are treated as pseudo-atoms) of our Transferable Potentials for Phase Equilibria (TraPPE) force field is based on a relatively simple functional form to calculate the potential energy, U, of the system (this form is essentially identical to those used in the popular AMBER (*39*), CHARMM (*40*), and OPLS (*41*) force fields)

$$U(\mathbf{r}_\mathcal{N}) = \sum_{i<j}^{\text{charges}} \frac{q_i q_j}{4\pi\epsilon_0 r_{ij}} + \sum_{i<j}^{\text{atoms}} 4\epsilon_{ij} \left[\left(\frac{\sigma_{ij}}{r_{ij}}\right)^{12} - \left(\frac{\sigma_{ij}}{r_{ij}}\right)^6 \right] \\ + \sum_{\text{dihedrals}} u_{\text{dih}}(\phi) + \sum_{\text{angles}} \frac{k_\theta}{2}(\theta - \theta_{\text{eq}})^2 \qquad (4)$$

where $\mathbf{r}_\mathcal{N}$, q, r, ϵ, σ, ϕ, and θ denote the set of all Cartesian positions, the charges, the pair separations, the Lennard-Jones well depths and diameters, the dihedral angles, and the bond bending angles, respectively. The electrostatic part sums over all (partial) charges which usually are centered at atomic sites. Lennard-Jones potentials are used for the van der Waals (vdW) interactions. Simple cosine expansions and harmonic potentials are used to govern dihedral motion and angle bending. Fixed bond lengths are used throughout the simulations because experience has shown that replacing flexible (chemical) bonds with rigid bonds (of fixed lengths) has no measurable effect on phase equilibria calculations (*42*).

In the design of the TraPPE force field we follow an "inside-out" approach, i.e. starting with the "bonded" 1–2 (bond length), 1–3 (angle) and 1–4 (dihedral) interactions. Accurate determination of these bonded interactions from molecular simulation is impossible. Thus experimental data (structures and vibrational frequencies) or quantum mechanical calculations on small molecules have to be used to obtain these parameters. The torsional potentials for 1–4 interactions are taken from the best available quantum mechanical calculations, there are no vdW and electrostatic contributions to 1–4 interactions. Once the potentials for the bonded interactions have been found, we come to the crux of the development of transferable potentials: the determination of the nonbonded vdW and electrostatic interactions. Unfortunately, at present neither the (partial) charges nor the vdW potentials can be obtained with sufficient accuracy using *ab initio* quantum mechanics (43,44) and we have to resort to empirical fitting procedures.

Calculated vapor-liquid coexistence curves (VLCC) and other phase equilibria are extremely sensitive to very small changes of the nonbonded force field parameters. Therefore, single-component VLCC offer a very attractive route to determine force field parameters. In 1993, we presented the first alkane force field (called SKS) that was based on fitting to VLCC of medium-length alkanes (35,36). This model allowed for the prediction of the critical points of high-molecular weight alkanes (35) for which no reliable experimental data are available. However, the SKS force field failed to give satisfactory results for short alkanes. More recently, we have proposed an improved parameterization (the TraPPE force field) that uses one additional force field parameter (allowing for different Lennard-Jones diameters of methyl and methylene groups) (38). Agreement with experiment is now more satisfactory for all alkane chain lengths. The VLCC of ethane, n-pentane, and n-octane are shown in Figure 1. Considering the relative simplicity of the force field, the results are encouraging. Figure 2 depicts the critical temperatures and normal boiling points of the linear alkanes as functions of chain length. While the agreement is good for the critical properties, the united-atom TraPPE force field consistently underestimates the boiling points (overestimates the vapor pressures). Finite-size scaling techniques were not used for the determination of the critical properties, but it is believed that finite-size errors are small for the system sizes studied here (38). The SKS and TraPPE-UA models have also been employed in studies of transport properties, agreement with experiment is rather satisfactory at higher temperatures but becomes worse as the triple point is approached (45,46). Thus the limitations of these united-atom force fields are already apparent. [An explicit-hydrogen alkane force field (including Lennard-Jones sites at the centers of C–H bonds) has also been developed in our group. This all-atom force field gives good vapor pressures over the entire range of temperatures and predicts critical and boiling temperatures with an accuracy of around 1%. It also improves the prediction of heats of vaporization and vapor-liquid Gibbs free energies of transfer for the linear alkanes. However, an all-atom force field increases the number of interaction sites by a factor of roughly 3 and thus the computational burden by an order of magnitude. (Chen, B.; Siepmann, J. I.; *J. Phys. Chem. B*,

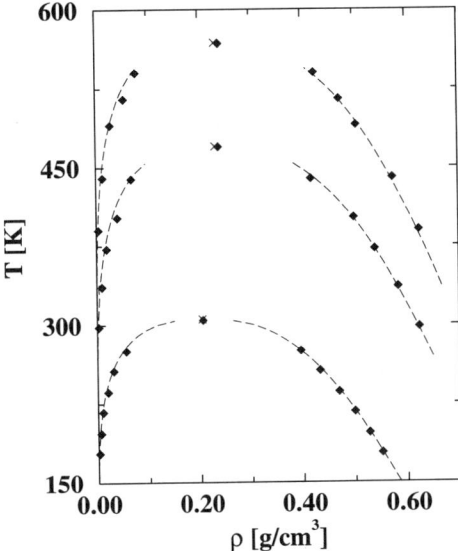

Figure 1. Vapor-liquid coexistence curves for ethane, n-pentane, and n-octane. Experimental coexistence data (57) and critical points (58) are shown as long dashed lines and crosses. Calculated saturated densities and extrapolated critical points for the TraPPE force field results are shown as diamonds.

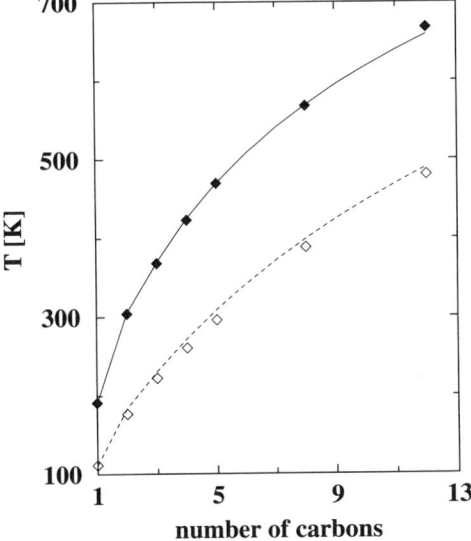

Figure 2. Critical temperatures (upper curves) and normal boiling points (lower curves). Experimental data (57,58) and simulation results for the TraPPE force field are shown as solid/dashed lines and filled/open diamonds, respectively.

submitted for publication)] Retaining the methyl and methylene parameters, we have also developed ternary and quarternary carbon parameters required for the modeling of branched alkanes (*30*). Figure 3 depicts the VLCC for 2,5- and 3,4-dimethylhexane. It is very encouraging, that without any special adjustable parameters the united-atom TraPPE force field is able to distinguish between these isomers. However, agreement with experiment is not as good as for the linear alkanes, in particular the critical temperatures are overestimated.

Multi-Component Phase Equilibria and Calculations of Gibbs Free Energies of Transfer

As a first step towards the simulation of chromatographic systems, we have carried out a detailed investigation of the partitioning of n-pentane and n-hexane between a helium vapor phase and a n-heptane liquid phase at standard conditions (*33,46*). This system was selected because of its similarity to GLC systems and of the availability of extremely accurate experimental partition data (*47*). Our initial studies (*33,46,48*) have demonstrated that configurational-bias Monte Carlo simulations in the isobaric (NpT) Gibbs ensemble are an efficient route to determine Gibbs free energies of transfer with the precision required for chromatographic studies. As discussed in detail by Schure (*49*), at room temperature, a dramatic change in the partitioning (or relative retention time), say by a factor 2, is associated with a relatively small change in free energy of $\Delta G° = 1.7$ kJ/mol. A 10% change in the partitioning requires only 0.23 kJ/mol. A reliable GEMC/CBMC simulation protocol (number of Monte Carlo cycles, system size, solute concentration, etc.) has been established (*46*). Inspired by experimental procedures, we have already demonstrated that the partitioning of multiple solutes can be obtained from one simulation (*33*). The precision of our calculations can also be enhanced by adjusting the phase ratio to yield roughly equal relative errors in the number densities in both phases (*33*).

Table 1 summarizes simulation details and results for n-pentane (solute or minor component)/n-heptane (liquid-phase solvent)/helium (carrier gas) systems. Neither replacing helium with argon as the carrier gas, nor increasing the system size from 350 to 1400 solvent plus carrier gas molecules, nor changing the solute mole fraction from 0.001 to 0.01 do significantly alter the partitioning of the alkanes. The standard errors of the mean for the Gibbs free energies of transfer are smaller than 0.4 kJ/mol, i.e. in the region required for quantitative predictions of chromatographic retention for high resolution separation systems (*49*). In comparison, the relative free energy of hydration for the conversion of methane to ethane obtained from molecular dynamics simulations has been reported to be 1.0 ± 0.8 kJ/mol (using FEP) and 2.2 ± 2.0 kJ/mol (using TI) (*14*). It should be noted here that the ΔG calculated for the united-atom TraPPE force field are approximately 1 kJ/mol smaller in magnitude than the corresponding experimental data. We attribute these differences to inaccuracies of the united-atom TraPPE force field. As noted above, the united-atom version yields consistently too high vapor pressures, while it gives very good results for

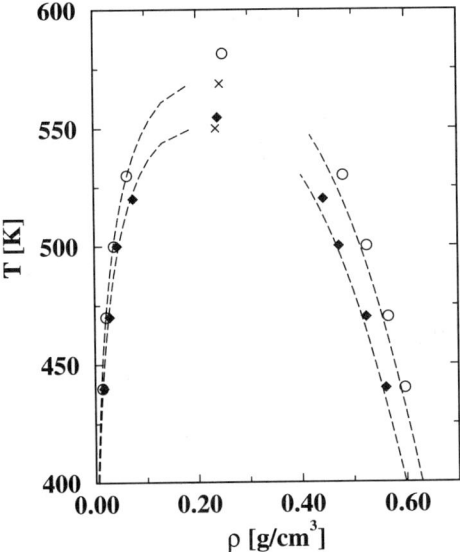

Figure 3. Vapor-liquid coexistence curves for 2,5-dimethylhexane and 3,4-dimethylhexane. Experimental coexistence data and critical points (57) are shown as long dashed lines and crosses. Calculated saturated densities and extrapolated critical points for the TraPPE force field results are shown as diamonds (2,5-DMH) and circles (3,4-DMH).

liquid densities and critical temperatures (38). Too high vapor pressures are consistent with too small ΔG. [The explicit-hydrogen version of the TraPPE force field (see above) does not suffer from this shortcoming and its use leads to much improved accuracies in the free energies of transfer.]

Table 1. Simulation details and results for ternary (helium or argon)/ n-heptane/n-pentane systems at a temperature of 298.15 K and a pressure of 101.3 kPa calculated for the TraPPE force field. N is the total number of atoms/molecules of a given type in the two simulation cells. ΔG and ρ denote Gibbs free energies of transfer and specific densities, respectively. The subscripts denote the statistical uncertainties in the simulation results. The experimental results are [46]: ΔG (C$_5$) = -13.7 kJ/mol and ΔG (C$_7$) = -19.5 kJ/mol.

		System A	System B	System C	System D
N (He)		210	0	840	840
N (Ar)		0	210	0	0
N (C$_5$)		2	2	8	1
N (C$_7$)		140	140	560	560
ΔG (He)	[kJ/mol]	7.48_{15}	–	7.56_{17}	7.53_{11}
ΔG (Ar)	[kJ/mol]	–	1.35_{15}	–	–
ΔG (C$_5$)	[kJ/mol]	-12.75_{24}	-12.64_{13}	-12.71_{20}	-12.52_{33}
ΔG (C$_7$)	[kJ/mol]	-18.24_{35}	-18.18_{9}	-18.26_{14}	-18.24_{22}
ρ_{liq}	[g/ml]	0.6845_{9}	0.6848_{12}	0.6843_{6}	0.6849_{6}
ρ_{vap}	[g/ml]	0.00060_{6}	0.00191_{9}	0.00060_{2}	0.00058_{4}

It is also important to test how the CBMC/GEMC methodology and the TraPPE force field perform at elevated temperatures and pressures such as those encountered in gas chromatography. To illustrate this, we have calculated the solubilities of helium in n-hexadecane over a wide range of temperatures and pressures (see Figure 4). Agreement with the experimental results is satisfactory over this range of conditions and also at standard conditions (Zhuravlev, N. D.; Siepmann, J. I.; 'Solubilities of He, Ar, CH$_4$, and CF$_4$ in normal alkanes with 6 to 16 carbon atoms: A Monte Carlo Study', in preparation).

Alkane Partitioning in a Helium/Squalane GLC System

Of all of the various chromatographic techniques, gas-liquid chromatography (GLC or just GC) is perhaps the simplest system to study on a fundamental level. This is due to the rather well-defined nature of a gas in equilibrium with a high boiling temperature liquid typically coated on the inner surface of a fused silica capillary. The retention of isomeric alkanes in GLC using high-molecular-weight branched alkanes as the liquid phase is the starting point in our endeavor to contribute to the understanding of chromatography retention processes. Secondary effects, such as adsorption at the silica-liquid interface and adsorption at the liquid film-vapor interface, are negligible in GLC separations involving solutes and liquid phase of very similar chemical nature (50,51).

Currently, we are investigating the partitioning of linear and branched alkanes with 5 to 8 carbon atoms between helium and squalane (2,6,10,15,19,23-hexamethyltetracosane), a widely used liquid phase in GLC (52) and the reference material for the Rohrschneider-McReynolds scheme of liquid phase characterization (53,54). During the GEMC simulations, particle swap moves have to be performed only for the solutes and the carrier gas. Since a liquid phase in GLC can only be used over a temperature range where its vapor pressures is negligible, there is no need to sample the partitioning of squalane. Two independent simulations are carried out that each sample the partitioning of a group of four solutes (n-pentane/2-methyl-pentane/3-methylpentane/n-hexane and n-heptane/2,5-dimethylhexane/3,4-dimethylhexane/n-octane). The simulation systems contain 96 squalane, a total of 8 solute (2 each) molecules and 200 (for pentane/hexane) or 500 (for heptane/octane) helium atoms. Preliminary estimates have been obtained for the solute partitionings and, at present, the statistical errors in the Gibbs free energies of transfer are around 0.5 kJ/mol. Experimental free energies of transfer are not available. Therefore, the simulation results are shown in the form of Kovats retention indices in Figure 5. The Kovats retention index I of a solute x (55,56) can be calculated directly from the solute partitionings using

$$I_x = 100n + 100 \left[\frac{\log(K_x/K_n)}{\log(K_{n+1}/K_n)} \right] . \quad (5)$$

where K_x, K_n, and K_{n+1} are the partition constants of the solute in question, the highest normal alkane (having n carbon atoms) that elutes prior to the solute, and the lowest normal alkane that elutes after the solute, respectively. It is encouraging that the elution order is predicted correctly for all eight solutes. However, the retention indices appear to be overestimated for all four branched alkanes.

Conclusions

The rapid development of more efficient simulation algorithms, more accurate force fields, and more powerful computers is now permitting the use of molecular simulations for the investigation of complex problems in thermodynamics that were hitherto intractable. The combination of Gibbs-ensemble Monte Carlo and configurational-bias Monte Carlo allows the efficient and precise determination of single and multicomponent phase diagrams and of Gibbs free energies of transfer. In conjunction with the united-atom TraPPE force field, results with

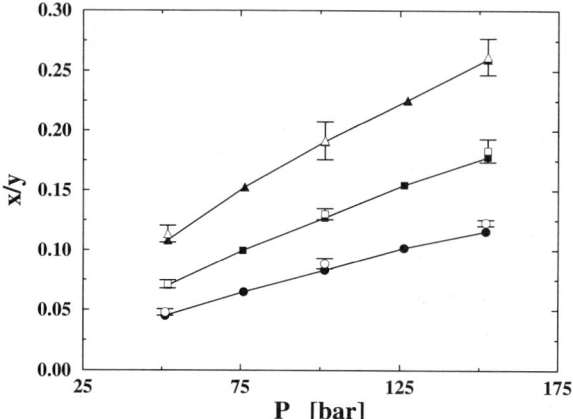

Figure 4. Solubility of helium in n-hexadecane as a function of temperature and pressure. x/y is the ratio of the mole fractions of helium in the liquid and vapor phases. The experimental data (59) and simulation results are depicted by filled and open symbols ($T = 464$ K: circles; $T = 545$ K: squares; $T = 624$ K: triangles).

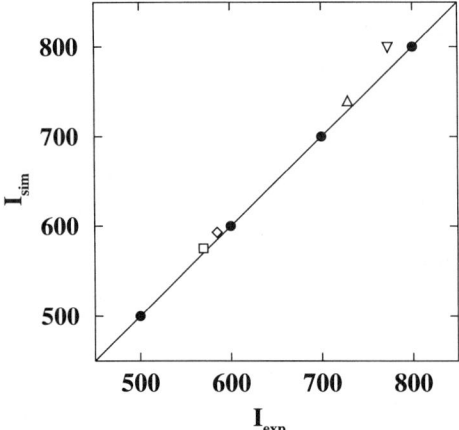

Figure 5. Predicted Kovats retention indices of four branched alkanes (2-methyl-pentane: square; 3-methylpentane: diamond; 2,5-dimethylhexane: triangle up; 3,4-di-methylhexane: triangle down) for a helium/squalane GLC system at $T = 343$ K. The experimental data were taken from (52).

a satisfactory degree of accuracy can be obtained, but further improvements in force fields are clearly desirable. The authors hope that quantitative molecular simulations of chromatographic systems will become an essential tool for providing a molecular-level understanding of the factors contributing to the retention process and for predicting retention times.

Acknowledgements. We would like to acknowledge many stimulating discussions with Pete Carr. Financial support from the National Science Foundation (CHE-9816328), the Petroleum Research Fund administered by the American Chemical Society, a Camille and Henry Dreyfus New Faculty Award, a McKnight/Land-Grant Professorship, and an Alfred P. Sloan Research Fellowship is gratefully acknowledged. MGM would like to thank the Department of Energy for a Computational Science Graduate Fellowship. A very generous amount of computer time was provided by the Minnesota Supercomputing Institute through the University of Minnesota–IBM Shared Research Project and NSF grant CDA-9502979.

Literature Cited
[1] Karger, B. L.; Snyder, L. R.; Eon, C.; *Anal. Chem.* **1978**, *50*, 2126–2136.
[2] Melander W. R.; Horváth, C.; in *High-Performance Liquid Chromatography: Advances and Perspectives*, Horváth, C., Ed.; Academic Press: London, 1980, Vol. 2, pp. 113–319.
[3] Martire, D. E.; Boehm, R. E.; *J. Phys. Chem.* **1983**, *87*, 1045–1062.
[4] Kaliszan, R.; *Quantitative Structure-Chromatographic Retention Relationships*, Chemical Analysis; Wiley: New York, 1987, Vol. 93.
[5] Lochmüller, C. H.; Reese, C.; Aschman, A. J.; Breiner, S. J.; *J. Chromatogr. A* **1993**, *656*, 3–18.
[6] Prausnitz, J. M.; Anderson, T. F.; Greens, E. A.; Eckert, C. A.; Hsieh, R.; O'Connell, J. P.; *Computer Calculations for Multicomponent Vapor-Liquid and Liquid-Liquid Equilibria*; Prentice-Hall: Englewood Cliffs, 1980.
[7] Smith, R. M.; *J. Chromatogr. A* **1993**, *656*, 381–415.
[8] Ben-Naim, A.; *Statistical Thermodynamics for Chemists and Biochemists*; Plenum Press: New York, 1992.
[9] Allen, M. P.; Tildesley, D. J.; *Computer Simulation of Liquids*; Oxford University Press: Oxford, 1987.
[10] Frenkel, D.; Smit, B.; *Understanding Molecular Simulation*; Academic Press: New York, 1996.
[11] Jorgensen, W. L.; *Acc. Chem. Res.* **1989**, *22*, 184–189.
[12] Frenkel, D.; in *Computer Simulation in Materials Science*, Meyer, M.; Pontikis, V.; Eds.; NATO ASI Ser. E205; Kluwer: Dordrecht, 1991; pp. 85–117.
[13] Kollman, P. A.; *Chem. Rev.* **1993**, *93*, 2395–2417.
[14] Radmer, R. J.; Kollman, P. A.; *J. Comp. Chem.* **1997**, *18*, 902–919.
[15] Chipot, C; Kollman, P. A.; Pearlman, D. A.; *J. Comp. Chem.* **1996**, *17*, 1112–1131.

[16] Kofke, D. A.; Glandt, E. D.; *Mol. Phys.* **1988**, *64*, 1105–1131.
[17] Valleau, J. P.; *J. Chem. Phys.* **1993**, **99**, 4718–4728.
[18] Kumar, S. K.; Szleifer, I.; Panagiotopoulos, A. Z.; *Phys. Rev. Lett.* **1991**, *66*, 2935–2938.
[19] Kofke, D. A.; *J. Chem. Phys.* **1993**, **98**, 4149–4162.
[20] Wilding, N. B.; *Phys. Rev. E* **1995**, **52**, 602–611.
[21] Escobedo, F. A.; de Pablo, J. J.; *J. Chem. Phys.* **1997**, *106*, 2911–2923.
[22] Spyriouni, T.; Economou, I. G.; Theodorou, D. N.; *Phys. Rev. Lett.* **1998**, *80*, 4466-4469.
[23] Panagiotopoulos, A. Z.; *Mol. Phys.* **1987**, *61*, 813–826.
[24] Panagiotopoulos, A. Z.; Quirke, N.; Stapleton, M.; Tildesley, D. J.; *Mol. Phys.* **1988**, *63*, 527–545.
[25] Smit, B.; de Smedt, P.; Frenkel, D.; *Mol. Phys.* **1989**, *68*, 931–950.
[26] Siepmann, J. I.; *Mol. Phys.* **1990**, *70*, 1145–1158.
[27] Siepmann, J. I.; Frenkel, D.; *Mol. Phys.* **1992**, *75*, 59–70.
[28] Frenkel, D.; Mooij, G. C. A. M.; Smit, B.; *J. Phys.: Cond. Matt.* **1992**, *4*, 3053–3076.
[29] de Pablo, J. J.; Laso, M.; Suter, U. W.; *J. Chem. Phys.*, **1992**, *96*, 2395–2403.
[30] Martin, M. G.; Siepmann, J. I.; *J. Phys. Chem. B* **1999**, in press.
[31] Mooij, G. C. A. M.; Frenkel, D.; Smit, B.; *J. Phys.: Cond. Matt.* **1992**, *4*, L255–L259.
[32] Laso, M.; de Pablo, J. J.; Suter, U. W.; *J. Chem. Phys.* **1992**, *97*. 2817–2819.
[33] Martin, M. G.; Siepmann, J. I. *J. Am. Chem. Soc.* **1997**, *119*, 8921–8924.
[34] Siepmann, J. I.; McDonald, I. R.; *Mol. Phys.* **1992**, *75*, 225–229.
[35] Siepmann, J. I.; Karaborni, S.; Smit, B.; *Nature*, **1993**, *365*, 330–332.
[36] Smit, B.; Karaborni, S.; Siepmann, J. I.; *J. Chem. Phys.* **1995**, *102*, 2126–2140; ibid. **1998**, *109*, 352.
[37] Siepmann, J. I.; Martin, M. G.; Mundy, C. J.; Klein, M. L.; *Mol. Phys.* **1997**, *90*, 687–693.
[38] Martin, M. G.; Siepmann, J. I.; *J. Phys. Chem. B* **1998**, *102*, 2569–2577.
[39] Cornell, W. D.; Cieplak, P.; Bayly, C.; Gould, I. R.; Merz, K. M.; Ferguson, D. M.; Spellmeyer, D. C.; Fox, T.; Caldwell, J. W.; Kollman, P. A.; *J. Am. Chem. Soc.* **1995**, *117*, 5179–5197.
[40] Mackerell, A. D.; Wiorkiewieczkuczera, J.; Karplus, M.; *J. Am. Chem. Soc.* **1995**, *117*, 11946–11975.
[41] Jorgensen, W. L.; Maxwell, D. S.; Tirado-Rives, J.; *J. Am. Chem. Soc.* **1996**, *118*, 11225–11236.
[42] Mundy, C. J.; Siepmann, J. I.; Klein, M. L.; *J. Chem. Phys.* **1995**, *102*, 3376–3380.
[42] Margenau, H.; Kestner, N.; *The Theory of Intermolecular Forces* (Pergamom Press: Oxford, 1970).
[43] Maitland, G. C.; Rigby, M.; Smith, E. B.; Wakeham, W. A.; *Intermolecular Forces: Their Origin and Determination*; Pergamon Press: Oxford, 1987.

[44] Mundy, C. J.; Klein, M. L.; Siepmann, J. I.; *J. Phys. Chem.* **1996**, *100*, 16779-16781.
[45] Mondello, M.; Grest, G. S.; Webb, E. B.; Peczak, P.; *J. Chem. Phys.* **1998**, *109*, 798-806.
[46] Martin, M. G.; Siepmann, J. I.; *Theo. Chem. Acc.* **1998**, *99*, 347-350.
[47] Eikens, D. I.; *Applicability of theoretical and semi-empirical models for predicting infinite dilution coefficients*; Ph.D. Dissertation, University of Minnesota, 1993.
[48] Martin, M. G.; Chen, B.; Zhuravlev, N. D.; Carr, P. W.; Siepmann, J. I.; *J. Phys. Chem. B* **1999,** in press.
[49] Schure, M. R.; in *Advances in Chromatography*, Brown, P. R.; Grushka, E.; Eds.; Marcel Dekker: New York, 1998, Vol. 39; pp. 139-200.
[50] Defayes, G.; Fritz, D. F.; Görner, T.; Huber, G.; de Reyff, C.; Kovats, E.; *J. Chromatogr.* **1990**, *500*, 139-184.
[51] Berezkin, V. G.; *Gas-Liquid-Solid Chromatography*; Marcel Dekker: New York, 1991.
[52] Tourres, D. A.; *J. Chromatogr.* **1967**, *30*, 357-377.
[53] Rohrschneider, L.; *J. Chromatogr.* **1966,** *22*, 6-22.
[54] McReynolds, W. O.; *J. Chromatogr. Sci.* **1970**, *8*, 685-691.
[55] Kovats, E.; *Helv. Chim. Acta* **1958,** *41*, 1915-1932.
[56] Kovats, E.; in *Advances in Chromatography*; Marcel Dekker: New York, 1965, Vol. 1; pp. 229-247.
[57] Smith, B. D.; Srivastava, R.; *Thermodynamic Data for Pure Compounds: Part A Hydrocarbons and Ketones*; Elsevier: Amsterdam, 1986.
[58] Teja, A. S.; Lee, R. J.; Rosenthal, D.; Anselme, M.; *Fluid Phase Equil.*, **1990**, *56*, 153.
[59] Lin, H. M.; Lee, R. J.; Lee, M. J.; *Fluid Phase Eq.* **1995,** *111*, 89-99.

Chapter 7

Stationary- and Mobile-Phase Interactions in Supercritical Fluid Chromatography

Sihua Xu, Phillip S. Wells, Yingmei Tao, Kwang S. Yun, and J. F. Parcher[1]

Chemistry Department, University of Mississippi, University, MS 38677

Mass spectrometric tracer pulse chromatography was used to measure the solubility of carbon dioxide in poly(dimethylsiloxane) over a wide range of pressures (15-100 atm) and temperatures from 35 to 120 °C. The data were compared with previously published results obtained with a wide variety of experimental methods including piezoelectric, gravimetric, dilatometric, and chromatographic procedures. The results are in agreement for pressures below the critical pressure of CO_2 but differ considerably for higher pressures. The lattice fluid model proposed by Sanchez and Lacombe, and later applied specifically to chromatographic systems by Martire and Boehm, was used to model the experimental solubility data using both measured characteristic parameters (ρ^*, T^* and P^*) and critical constants as reduction parameters for the model. The models were used to calculate the interaction parameter, χ, from the experimental data. The results showed that the interaction parameter varied inversely with temperature as expected; however, the measured parameter also changed with the composition of the CO_2-polymer mixture. This composition dependence is not predicted from the model and indicates that the models as currently structured are not strictly applicable to the CO_2-PDMS system at temperatures and pressures close to critical.

With the advent of supercritical fluid chromatography, SFC, the apparent gap between gas and liquid chromatography was finally occupied by a new chromatographic technique in which the mobile phase could vary continuously from a gas at low pressure

[1]Corresponding author.

to a supercritical fluid or even a liquid at high pressures (*1*). This integration of experimental techniques led to the concept of unified chromatography promulgated by Giddings (*2*). The idea of unified chromatography has not been achieved in practice for a variety of reasons including instrumentation problems, the wide disparity of sample types, and perhaps most importantly the lack of an all-encompassing theoretical foundation. The latter problem has been addressed by Martire in a collection of papers developing the concept of a "Unified Molecular Theory of Chromatography" (*3-7*). Another series of theoretical treatments for SFC in particular has been proposed by Roth (*8-13*). These statistical thermodynamic treatments cover a wide variety of mobile and stationary phases but suffer from a common lack of appropriate experimental data required for verification. Fundamental thermodynamic data are particularly difficult to measure for systems, such as those often encountered with SFC, in which an exact delineation between the stationary and liquid phases is difficult. Thus, experimental measurement of the column void volume (mobile phase volume) and the volume as well as the composition of the stationary phase is often difficult if not impossible. Nevertheless, these crucial parameters are absolutely necessary for the verification of any proposed statistical thermodynamic model.

The uncertainty in the volumes and compositions of the stationary and mobile phases arises because the mobile phase, usually CO_2, dissolves in or adsorbs on most stationary phases. In most lattice fluid SFC models, the "composition" of the mobile phase, which is usually pure CO_2, is also treated as a variable because the mobile phase is considered to be a binary mixture of CO_2 molecules and unoccupied space, vacancies, or empty lattice sites. Density thus becomes the composition variable for the mobile phase.

Using this definition, the composition of both the stationary and mobile phases varies significantly with temperature and pressure, especially near the critical point of the mobile phase. At temperatures below critical, the composition of the mobile phase may change discontinuously from the density of a gas to that of a liquid or *vice versa*. Thus, supercritical fluid chromatographic systems are very complex even in the absence of chromatographic solutes, modifiers, or additives. The influence of such a dissolution or adsorption process on the retention and resolution of any chromatographic solutes is uncertain and difficult to ascertain, and is thus most often simply ignored. Fortunately, such disregard seems justifiable in view of the remarkable results obtained with very complex SFC systems.

Most SFC models consider the mobile phase composition (density) to be the controlling factor for the retention of typical solutes. This makes the SFC models consistent with HPLC models although the SFC mobile phase composition is now a function of temperature as well as pressure. A complicating factor, however, is the fact that many engineering studies have shown that CO_2 dissolves extensively in common polymeric stationary phases. The most thoroughly studied fluid-polymer system closely related to any practical SFC setup is CO_2 with PDMS, poly(dimethylsiloxane). Several disparate experimental techniques have been used to measure the high-pressure solubility isotherms of this system. In 1986, Fleming and Koros (*14*) used a pressure decay cell to measure the uptake of CO_2 by silicone rubber at 35 °C. The sorption measurements required a correction to the void volume of the cell due to the swelling of the CO_2-impregnated polymer. The dilation of the polymer was measured with a cathotometer. In a later study, Briscoe and Zakaria (15) measured the absorption of CO_2 at 42 °C from changes in the resonance frequency of a metal beam coated with PDMS. The dilation was measured by an ultrasonic technique. In 1994, Garg, et al.(16), used a gravimetric method to measure the solubility of CO_2 in PDMS at 50, 80, and 100 °C. Recently, a chromatographic method (19) was used to investigate the same systems at 30, 40, and 50 °C. Two additional studies reported dilation data at 41 °C (17) and 35, 50, and 70 °C (18). This variety of experimental methods, viz., gravimetric, barometric, ultrasonic, and chromatographic techniques, provides a reliable solubility database which is remarkably consistent *at pressures below 80 atm.* An example of the experimental results at 40 °C is shown in Figure 1. Below 80 atm, the solubility data obtained from a wide variety of experimental techniques agree; however, at higher pressures there are obvious discrepancies which are almost certainly not due to experimental errors.

Other disparities appear in the measured partial molar volumes of carbon dioxide which can be measured from experiments (14, 15, 20) where both sorption and swelling data were obtained. The partial molar volume, $\tilde{v}_1$, is defined as

$$\tilde{v}_1 \equiv \left(\frac{\partial V}{\partial n_1} \right)_{T,P,n_2} \tag{1}$$

where V is the total volume of the swollen polymer and n_i represent the number of moles of component i in the mixture. This expression can be transformed to a more useful form for experimental determination of $\tilde{v}_1$, *viz.,*

$$\tilde{v}_1 = M_1 \left[v + w_2 \left(\frac{\partial v}{\partial w_1} \right)_{T,P} \right] \tag{2}$$

M_1 is the molar mass of component 1, w_i is the weight fraction of component i in the mixture, and v is the specific volume of the mixture. Thus, $\tilde{v}_1$ can be determined directly from swelling, v, and sorption, w_1, experimental data.

A third form of the equation can be derived if hydrostatic compression of the polymer is taken into account. That is

$$\tilde{v}_1 = M_1 \left(v + w_2 \left[\left(\frac{\partial v}{\partial w_1} \right)_T + v\beta \left(\frac{\partial P}{\partial w_1} \right)_T \right] \right) \tag{3}$$

where β is the isothermal compressibility of the polymer. Each of these three equations was used to calculate $\tilde{v}_1$ from the experimental sorption and swelling data (14, 15, 20). The results are shown in Figure 2. Again, there is agreement between the different data

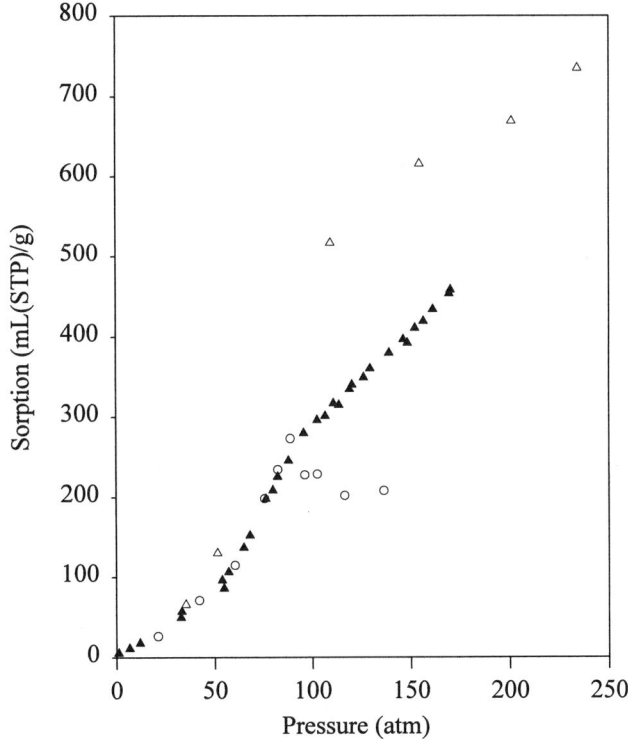

Figure 1 Literature Data for the Solubility of CO_2 in PDMS at ca. 40 °C

- ○ Tracer Pulse Chromatography (19)
- ▲ Calculated from Dilatometric data (17) assuming a Constant Molar Volume of CO_2 (46.2mL/mol)
- △ Resonance Frequency Method (15)

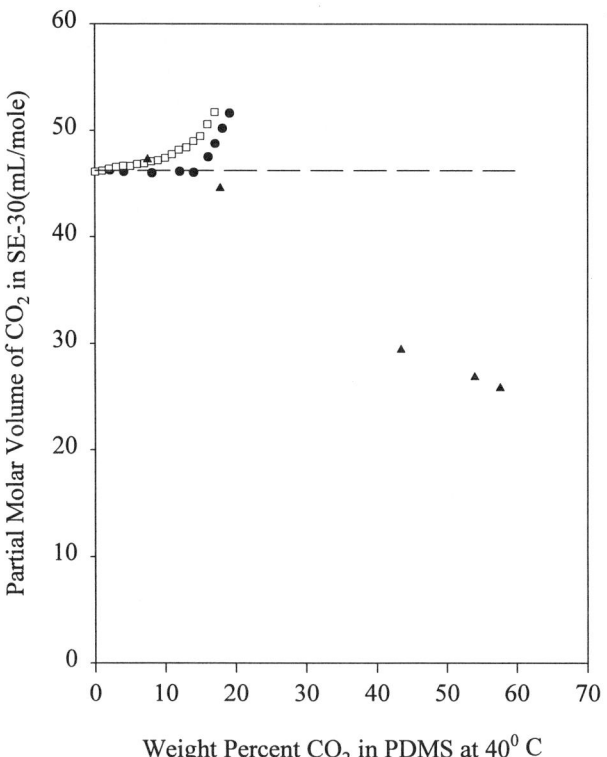

Figure 2 Experimental Values for the Partial Molar Volume of CO_2 in PDMS

- ● Fleming and Koros (*14*) (Equation 2)
- □ Pope, Sanchez, Koros, and Fleming (*20*) (Equation 3)
- ▲ Briscoe and Zakaria (*15*) (Equation 1)

sets at low concentration of CO_2 but significant deviation at higher concentrations.

If carbon dioxide dissolves in PDMS under chromatographic conditions to the extent illustrated by the upper curves in Figure 1, the effects of CO_2 dissolved in the stationary phase on the retention and resolution of SFC solutes should be readily apparent. Conversely, if CO_2 dissolves only to the extent illustrated by the lower curve in Figure 1, then this could explain the success of models based solely on mobile phase composition (density) as the controlling factor in SFC. The exact cause(s) of the observed solubility discrepancies is not known; however, it is clear that a critical need exists for further experimental investigations of systems, such as CO_2-PDMS, which can serve as excellent models for the very complex, multicomponent systems encountered in chromatography. In the present investigation, a chromatographic method, *viz.*, tracer pulse chromatography, was used to determine the solubility of CO_2 in PDMS (SE-30) over a wide range of pressure (to 100 atm) and temperature (35 -120 °C). The objective of this work is to provide a reliable set of data for the evaluation of current theoretical models for chromatographic systems.

Experimental

Instrumentation. The instrumentation and experimental procedures have been described in detail previously (*21*, *22*). Briefly, a closed-loop injection system containing neon and isotopically labeled CO_2 was used to make repetitive injections over a range of pressures and temperatures. A mass specific detector (HP 5971 MSD) was used to distinguish the isotopic carbon dioxide from the high background of unlabeled CO_2 carrier gas. SFC-grade natural carbon dioxide was obtained from Scott Specialty Gases. Isotopically labeled carbon dioxide ($^{13}C^{18}O^{16}O$) was purchased from Isotec. Neon was used as the dead time marker.

Column Material. The polymeric material used in this study was SE-30 (poly(dimethylsiloxane)) obtained from Supelco Chromatography Products. The molecular weight was approximately 100,000. The analytical columns were packed in this laboratory. Chromosorb W HP (60/80) with a surface area of 1 m^2/g was used as the solid support. The liquid loading was determined by solvent extraction with chloroform. The percent coating of PDMS was 18.35. This gave an average film thickness of 0.23 μm. The thin film allowed rapid equilibration of the CO_2-polymer systems.

Calculations. The total amount of CO_2 in the packed column, $n_{CO_2}^{Total}$ was determined from the retention time, t_R^*, of the isotopically labeled CO_2 and the measured molar flow rate, F_m, of CO_2 through the column from the relation: $n_{CO_2}^{Total} = F_m t_R^*$. Likewise, the amount of CO_2 in the mobile phase was determined from the retention time of neon, t_0, which was not measurably soluble in PDMS, *i.e.*, $n_{CO_2}^M = F_m t_0$. The amount of CO_2

dissolved in the polymer was determined by difference. The results are expressed in terms of mL(STP)/g of polymer to conform with previously published data.

The density of the pure fluid phase was calculated from the Ely equation-of-state (23). The density of PDMS was calculated at the experimental temperatures from the relation given by Shih and Flory (24).

$$\rho = 0.9919 - 8.92 x 10^{-4} t + 2.65 x 10^{-7} t^2 - 3 x 10^{-11} t^3$$

Results and Discussion

The experimental results for the solubility measurements are given in Table I for pressures up to 100 atm over the temperature range from 35 to 120 °C. The data are illustrated as solubility isotherms in Figure 3. Comparison of the chromatographic data with previously published solubility data obtained with other experimental procedures is illustrated in Figures 4-7. The data at low temperatures (T <80 °C) are remarkably consistent especially considering the wide variety of experimental techniques, *viz.*, gravimetric, dilatometric, chromatographic, and sonic methods, used for the solubility measurements. There is some discrepancy at higher temperatures (Figure 7); however, the agreement at lower temperatures is quite satisfactory.

Table I. Experimental Data for the Solubility of CO_2 in Poly(dimethylsiloxane).

Pressure (atm)	Amount of CO_2 Absorbed (mL(STP)/g polymer)					
	Temperature (°C)					
	35	40	50	80	100	120
15			16.4			6.09
20	29.0	26.3	21.6	15.3	13.1	12.1
30	49.0	44.4	36.3	24.7	18.8	14.3
40	71.3	63.0	50.5	33.0	27.5	20.4
50	95.8	83.6	68.5	43.6	35.6	28.6
60	132	117	90.6	57.9	44.7	35.0
70	172	149	111	61.9	49.9	40.4
80	214	196	131	78.8	58.4	46.3
90	242.	224	162	87.3	67.2	54.4
100	207	240	174	96.7	75.6	61.1

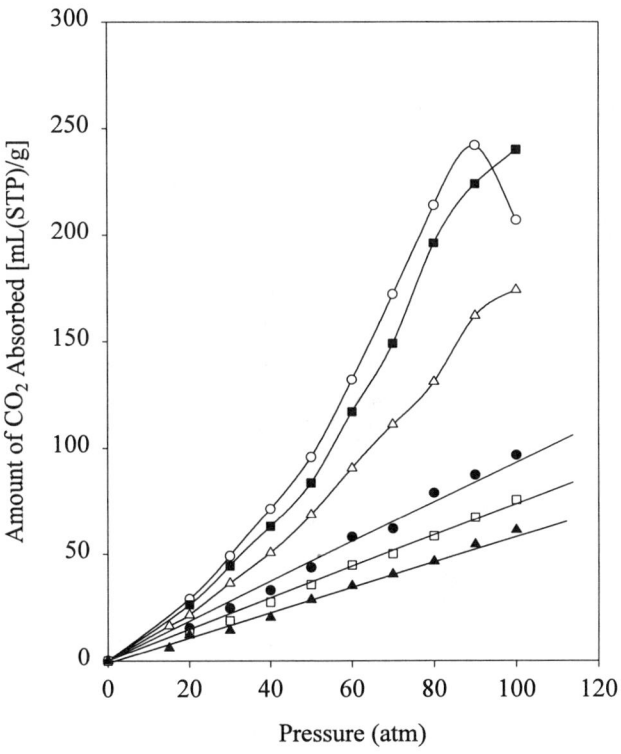

Figure 3 Absorption Isotherm for CO_2 in PDMS

○	35 °C	■	40 °C
△	50 °C	●	80 °C
□	100 °C	▲	120 °C

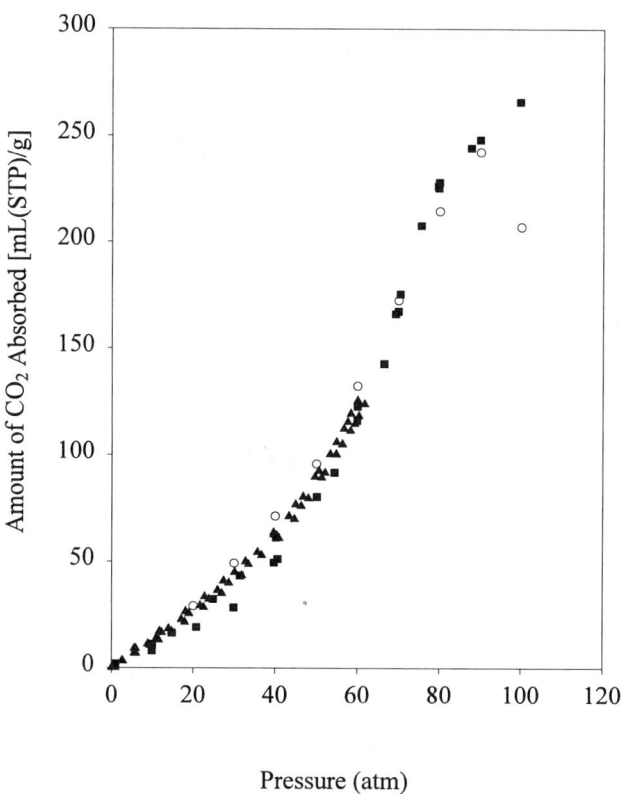

Figure 4 Experimental Isotherms at 35 ^{0}C

    ○   This work (TPC)
    ■   Shim and Johnston (Dilatometric) *(18)*
    ▲   Fleming and Koros (Barometric) *(14)*

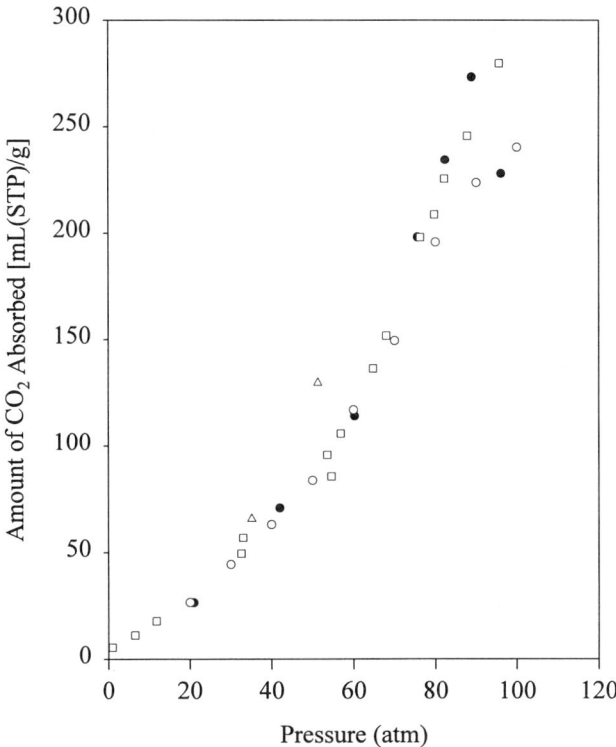

Figure 5 Experimental Isotherms at 40 ^{0}C

- ● Strubinger *et al.* (TPC) *(19)*
- □ Eckert and co-workers (Dilatometric) *(17)*
- △ Briscoe and Zakaria (Piezoelectric) *(15)*
- ○ This work (TPC)

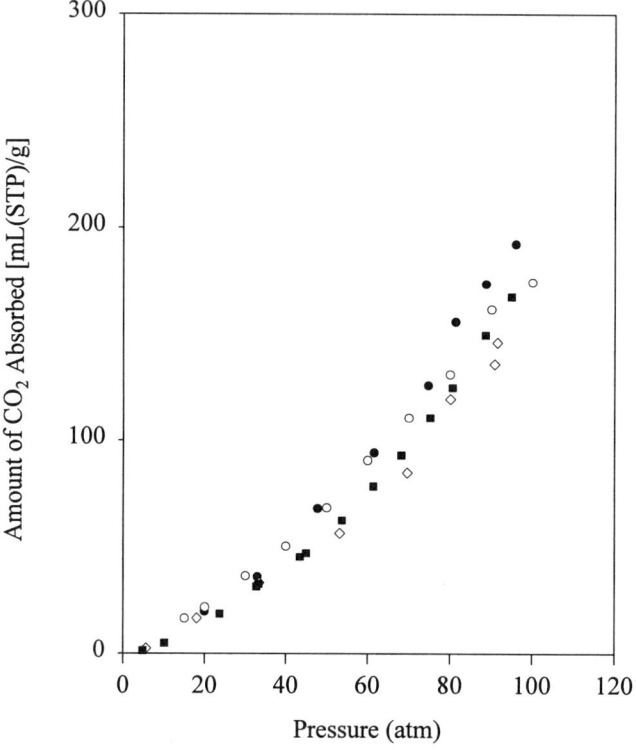

Figure 6 Experimental Isotherms at 50 °C

- • Strubinger *et al*. (TPC) *(19)*
- ◇ Garg *et al*. (Gravimetric) *(16)*
- ○ This work (TPC)
- ■ Shim and Johnson (Dilatometric) *(18)*

107

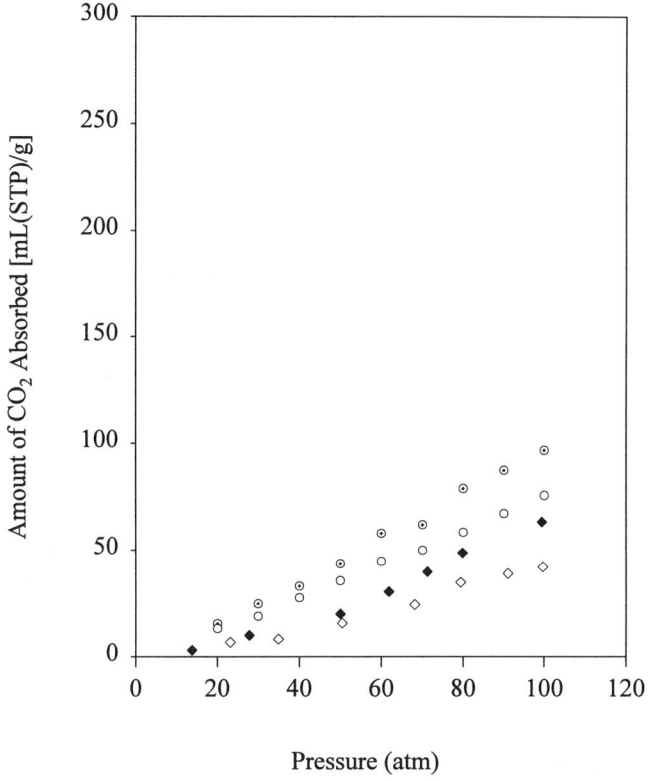

Figure 7 Experimental Isotherms at 80 °C and 100 °C

- ◆ Garg *et al.* (Gravimetric) 80 °C *(16)*
- ⊙ This work (TPC) 80 °C
- ◇ Garg *et al.* (Gravimetric) 100 °C *(16)*
- ○ This work (TPC) 100 °C

Theoretical Models

The extensive set of solubility data for CO_2 in PDMS provides an excellent database for the evaluation of existing models relating to the solubility of a compressible fluid in a polymer. The classical model for such systems is the Flory-Huggins equation for non-athermal systems.

Flory-Huggins Model This model, which is commonly used to interpret data for the solubility of gases in polymers, was proposed concurrently by Flory (25) and Huggins (26). The basic equation for this theory relates the activity of the gas, in this case CO_2 (component 1), in the polymer (component 2) to the volume fraction the polymer, ϕ_2, and an interaction parameter, χ_{12}.

$$\ln A_1 = \ln(1-\phi_2) + \left(1 - \frac{r_1}{r_2}\right)\phi_2 + \chi_{12}\phi_2^2 \quad (4)$$

where r_i is a size parameter for component i. In the case where component 1 is a gas and component 2 is a polymer, it is usually assumed that $\frac{r_1}{r_2} \ll 1$. The activity of CO_2 in the polymer can be calculated from the relation

$$A_1 = \frac{\gamma_1 P_1}{\gamma_1^{sat} P_1^0 \exp\left(\frac{\tilde{v}_1(P - P_1^0)}{RT}\right)} \quad (5)$$

where P_1 is the partial pressure of CO_2, P_1^0 is the saturated vapor pressure of pure, liquid CO_2 at temperature T, and γ_1 and γ_1^{sat} are the fugacity coefficients for CO_2 at T, P_1 and T, P_1^0, respectively. The denominator of equation 5 represents the fugacity of CO_2 in the standard state which is defined to be pure, liquid CO_2 at T and P. This is a commonly adopted standard state for liquids at $P > P^0$ and $T < T_c$. For CO_2, however, this standard state is awkward for two reasons. Firstly, pure liquid CO_2 does not exist at temperatures greater than the critical temperature, $T_c = 31\,°C$, so a van't Hoff-type extrapolation is required for $T > T_c$. Secondly, at pressures less than P^0, the Poynting correction, i.e., the exponential term, is questionable because $\tilde{v}_1$ represents the molar volume of *liquid* CO_2. Despite these problems, equation 5 is the most commonly used expression for calculating the activity of CO_2 in polymers. Figure 8 shows a plot of the logarithm of the activity coefficient of CO_2 in PDMS as a function of the volume fraction of polymer in the CO_2-PDMS mixture. Using activity as the determinant parameter effectively eliminates the temperature dependence of the solubility data.

In order to determine the interaction parameter, χ_{12} was used as the adjustable parameter. That is, for each of the data points given in Table I, the Flory-Huggins interaction parameter was calculated from equation 2, and the results are shown in Figure 9. The χ_{12} values average 1.0, but significant scatter is observed at low CO_2

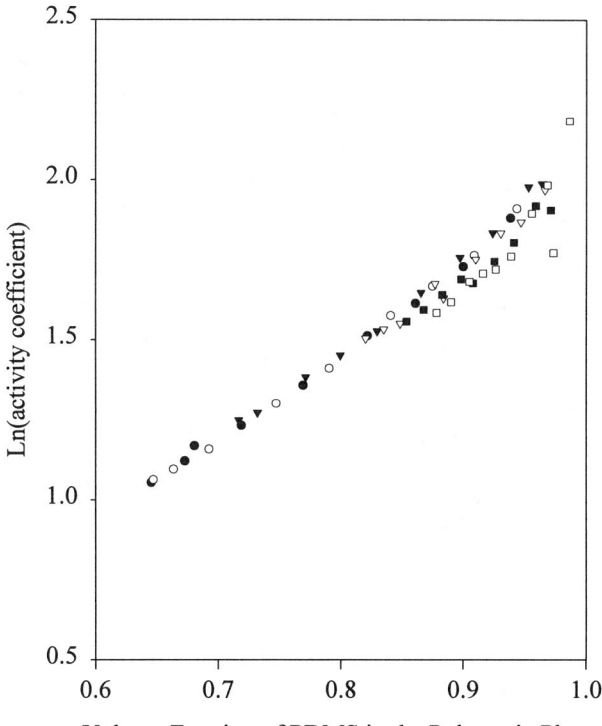

Figure 8 Plot of Activity Coefficient of CO_2 in PDMS as a Function of the Volume Fraction of PDMS

- ●   35 ^{0}C   ○   40 ^{0}C
- ▼   50 ^{0}C   ▽   80 ^{0}C
- ■   100 ^{0}C   □   120 ^{0}C

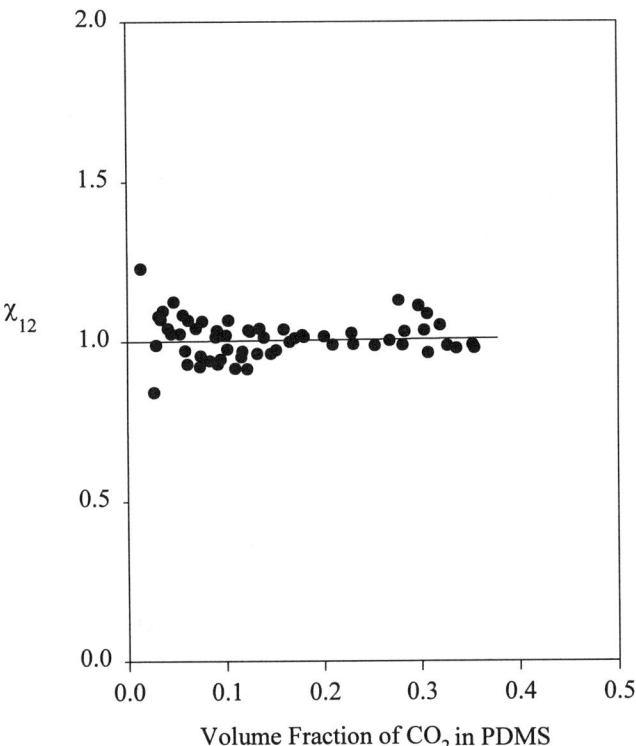

Figure 9. Flory-Huggins Interaction Parameter for the Experimental Data (Table I)

concentrations.

Sanchez and Lacombe Model Sanchez and co-workers (27-30) have proposed a series of lattice fluid models for fluid-polymer systems. The authors proposed an equation for the chemical potential of any component in an equilibrium mixture of compressible fluids dissolved in a polymer. The general equation is:

$$\frac{\mu_a}{RT} = \ln\phi_a\tilde{\rho} + 1 - r_a\sum_{i=1}^{k}\frac{\phi_i}{r_i} + r_a\tilde{\rho}\left\{\sum_{i=1}^{k}\phi_i\chi_{ai} - \sum_{j=1}^{j-1}\sum_{i=1}^{k}\phi_i\phi_j\chi_{ij}\right\} + r_a\left[\frac{-\tilde{\rho}}{\tilde{T}_a} + \frac{\tilde{P}_a}{\tilde{T}_a\tilde{\rho}} + \left(\frac{1-\tilde{\rho}}{\tilde{\rho}}\right)\ln(1-\tilde{\rho})\right] \quad (6)$$

where μ_a is the chemical potential of component a in a k component mixture. Two distinct concentration scales are used in the model. These are defined in terms of the size parameter, r_i, which describes the number of lattice sites occupied by a mole of component i; the number of moles of i, N_i; and N_0, the number of empty lattice sites in the fluid or condensed (polymer) phase. The concentration scales are

the fraction of occupied sites: $\quad \tilde{\rho} = \dfrac{\sum_{j=1}^{k}N_j r_j}{N_0 + \sum_{j=1}^{k}N_j r_j} \quad$ and

the volume fraction of component i: $\quad \phi_i = \dfrac{N_i r_i}{\sum_{j=1}^{k}N_j r_j}$

The product of these two concentrations, i.e., $\phi_i\tilde{\rho}$ gives the fraction of sites occupied by component i. The reduced equation-of-state parameters, $\tilde{\rho}, \tilde{T}_a$, and $\tilde{P}_a$ are defined in terms of the characteristic parameters ρ^*, T_a^*, and P_a^* by the relations $\tilde{\rho} = \dfrac{\rho}{\rho^*}$, $\tilde{T}_a = \dfrac{T}{T_a^*}$, $\tilde{P}_a = \dfrac{P}{P_a^*}$. The interaction between the species i and j in the solution is described by the interaction parameter $\chi_{ij} = \dfrac{1}{RT}(\varepsilon_{ii}^* + \varepsilon_{jj}^* - 2\varepsilon_{ij}^*)$ where ε_{ij}^* represents the total interaction energy between mers of components i and j.

A simple isotherm model for the dissolution of CO_2 (component 1) in a polymer, such as PDMS, (component 2) can be derived by equating the chemical potential of the CO_2 in the fluid (mobile) phase and the polymer (stationary phase) phase, i.e., $\dfrac{\mu_{1(s)}}{RT} = \dfrac{\mu_{1(m)}}{RT}$. For the mobile phase composed of pure CO_2, the chemical potential is

$$\frac{\mu_{1(m)}}{RT} = \ln\tilde{\rho}_m + r_1\left[\frac{-\tilde{\rho}_m}{\tilde{T}_1} + \frac{\tilde{P}_1}{\tilde{T}_1\tilde{\rho}_m} + \left(\frac{1-\tilde{\rho}_m}{\tilde{\rho}_m}\right)\ln(1-\tilde{\rho}_m)\right] \quad (7)$$

The stationary phase may contain both components 1 and 2, but component 2 is not volatile ($\phi_{2(m)} = 0$), so the chemical potential of component 1 in the stationary phase is given by

$$\frac{\mu_1(s)}{RT} = \ln\phi_{1(s)}\tilde{\rho}_s + 1 - r_1\left\{\frac{\phi_1(s)}{r_1} + \frac{\phi_2(s)}{r_2}\right\}$$
$$+ r_1\tilde{\rho}_s\left[\phi_{2(s)}\chi_{12} - \phi_{2(s)}\phi_{1(s)}\chi_{12}\right] + r_1\left[\frac{-\tilde{\rho}_s}{\tilde{T}_1} + \frac{\tilde{P}_1}{\tilde{T}_1\tilde{\rho}_s} + \left(\frac{1-\tilde{\rho}_s}{\tilde{\rho}_s}\right)\ln(1-\tilde{\rho}_s)\right] \quad (8)$$

The equation-of-state for either phase is given by the relation:

$$\frac{\tilde{P}}{\tilde{T}} + \frac{\tilde{\rho}^2}{\tilde{T}} + \ln(1-\tilde{\rho}) + \tilde{\rho}\left(1 - \frac{1}{r}\right) = 0 \quad (9)$$

If the chemical potentials are set equal, it is assumed that there are no empty lattices sites in the condensed (stationary) phase ($\tilde{\rho}_s = 1$), and the EOS for each phase is used to eliminate the last term of equations 7 and 8, the result is a form of isotherm equation:

$$\ln\phi_{1(s)} + 1 - r_1\left\{\frac{\phi_{1(s)}}{r_1} + \frac{\phi_{2(s)}}{r_2}\right\} + r_1\left\{\phi_{2(s)}\chi_{12} - \phi_{2(s)}\phi_{1(s)}\chi_{12}\right\} - \frac{r_1}{\tilde{T}_1} =$$
$$\ln\tilde{\rho}_m + r_1\left[-\frac{\tilde{\rho}_m}{\tilde{T}_1} - \frac{\tilde{\rho}_m}{\tilde{T}_1} - \left(1 - \frac{1}{r_1}\right) + \frac{\tilde{\rho}_m^2}{\tilde{T}_1} + \tilde{\rho}_m\left(1 - \frac{1}{r_1}\right)\right] \quad (10)$$

Substituting $\phi_{2(s)} = 1 - \phi_{1(s)}$ and rearranging gives the final isotherm equation for the Sanchez-Lacombe lattice fluid model.

$$\ln\phi_{1(s)} + r_1\left\{\frac{1}{r_2} - \frac{1}{r_1}\right\}\phi_{1(s)} + r_1\chi_{12}(1-\phi_{1(s)})^2 = \ln\tilde{\rho}_m + r_1\left(1 - \frac{1}{r_1}\right)\tilde{\rho}_m + \frac{r_1}{\tilde{T}_1}(1-\tilde{\rho}_m)^2 + r_1\left(\frac{1}{r_2} - 1\right) \quad (11)$$

In order to test this model with experimental data, it is necessary to convert the solubility data given in Table I to the volume fraction of CO_2 in the polymeric phase. This requires a knowledge of the partial molar volume, $\tilde{v}_1$, of CO_2 in the polymer phase at the experimental temperatures and pressures. If $\tilde{v}_1$ is known, the volume fraction can be calculated from the expression
$$\phi_{1(s)} = \frac{\frac{S\tilde{v}_1}{22,400}}{\frac{S\tilde{v}_1}{22,400} + \frac{1}{\rho_2}}$$
where S is the amount of CO_2 absorbed in units of mL(STP)/g of polymer and ρ_2 is the density (g/mL) of the polymer at the experimental temperature. For comparison, a constant value of $\tilde{v}_1 = 46.2$ mL/mole was assumed (dashed line in Figure 2).

The second step in fitting the model to the experimental data involves the determination of the size parameter, r_1, and the characteristic parameters ρ_1^* and T_1^* for carbon dioxide. Several sets of these equation-of-state parameters for CO_2 have been published previously; however, the values vary depending upon the type of pure component data used to determine the parameters. In this work, the parameters were determined from the PVT data for pure CO_2 using the Ely (23, 31) equation-of-state for pure CO_2. The results are given in Table II along with literature results. The fit of the Sanchez-Lacombe equation-of-state to the data generated from the Ely EOS are shown in Figure 10.

Table II. Characteristic Parameters for Pure Carbon Dioxide

Reference	T_1^*	P_1^*	ρ_1^*	r_1
Sanchez and Panayiotou (30)	283	6510	1.62	7.6
Pope, et al. (20)	283	6510	1.62	7.6
Condo, et al (33)	308	5670	1.51	6.5
Garg, et al (16) Pressure Optimization	328	4582	1.43	5.2
Density Optimization	330	4520	1.43	5.1
This Work	311	5056	1.55	6.0

The isotherm model represented by equation 11 was fit to the experimental data given in Table I with the size and characteristic parameters given in the last row of Table II with only the interaction parameter, χ_{12}, used as a mixture (adjustable) parameter. The results are illustrated in Figure 11 which shows the calculated χ_{12} values as a function of temperature and composition of the CO_2-polymer mixture. The interaction parameter increases with 1/T indicating that $\frac{\varepsilon_{11}^* + \varepsilon_{22}^*}{2} > \varepsilon_{12}^*$. This behavior is reasonable; however, the composition dependence of the calculated χ_{12} values casts doubt on the validity of the Sanchez-Lacombe isotherm model as expressed by equation 11.

Martire and Boehm Model In 1987, Martire and Boehm (4) presented a lattice fluid model for gas- liquid- and supercritical fluid chromatography. This "Unified Molecular Theory of Chromatography" was developed in the form of a series of treatments for specific chromatographic systems. The theoretical approach was based on the Sanchez-Lacombe lattice fluid model; however, only one concentration scale was used. That is the "volume fraction" scale defined by the relation $\theta_i = \frac{N_i r_i}{N_0 + \sum_{j=1}^{k} N_j r_j}$. This quantity represents the fraction of all lattice sites occupied by component i and is functionally equivalent to Sanchez-Lacombe's product $\phi_i \tilde{\rho}$, i.e., the fraction of sites occupied by component i. In the case of a pure CO_2 mobile phase, $\theta_{1(m)} = \phi_{1(m)} \tilde{\rho}_{(m)} = \tilde{\rho}_{(m)}$. For a condensed stationary

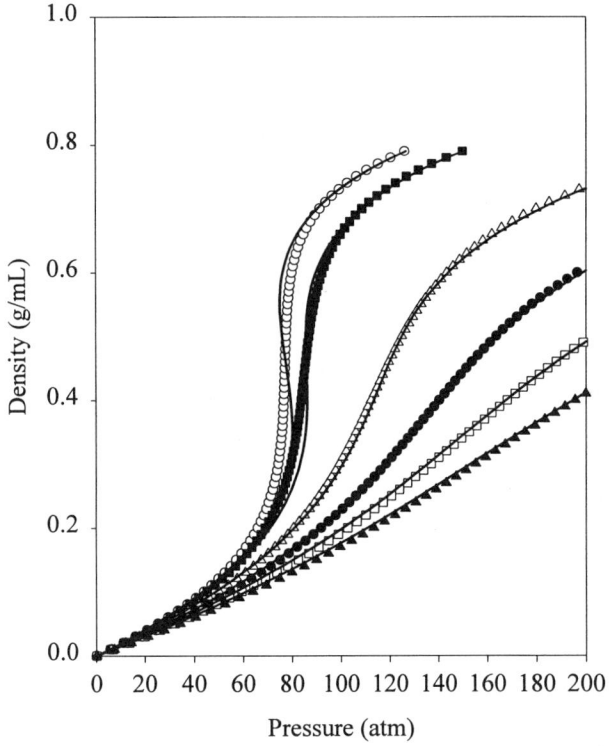

Figure 10 Equation-of-State Calculations for the Density of CO_2

Line	Sanchez-Lacombe Equation-of-State (Equation 9)
Symbols	Ely Equation-of- State

○	35 °C	■	40 °C
△	60 °C	●	80 °C
□	100 °C	▲	120 °C

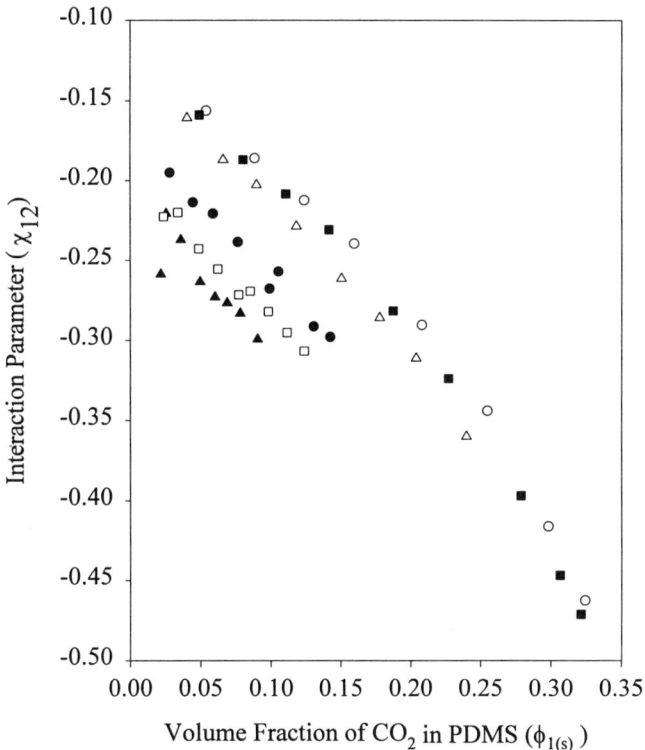

Figure 11 Plot of Interaction Parameter as a Function of the Volume Fraction of CO_2 in PDMS

- ○ 35 °C
- △ 50 °C
- □ 100 °C
- ■ 40 °C
- ● 80 °C
- ▲ 120 °C

phase $\tilde{\rho}_s = 1$, so $\theta_{1(s)} = \phi_{1(s)}\tilde{\rho}_{(s)} = \phi_{1(s)}$. A second difference between the two models concerns the definition of the interaction terms, ε and χ. Martire defined the interaction parameter $\chi_{ij} = \dfrac{-z}{2RT}(\varepsilon_{ii} + \varepsilon_{jj} - 2\varepsilon_{ij})$. The relation between the interaction energies is given by $\varepsilon^*_{ii} = \dfrac{-z}{2}\varepsilon_{ii}$ where ε^*_{ii} is the total interaction energy per mer, z is the coordination number, and ε_{ii} is the nonbonded, mer-mer interaction energy. Martire assumed $\varepsilon_{ii} < 0$ to indicate attractive forces between mers. Finally, the reduced temperature is given by $\dfrac{1}{\tilde{T}_1} = -\dfrac{z\varepsilon_{11}}{2RT}$. Using these relations, the isotherm equation 11 becomes

$$\ln\theta_{1(s)} + r_1\left\{\dfrac{1}{r_2} - \dfrac{1}{r_1}\right\}\theta_{1(s)} + r_1\chi_{12}(1-\theta_{1(s)})^2 = \ln\theta_{1(m)} + r_1\left(1 - \dfrac{1}{r_1}\right)\theta_{1(m)} - \dfrac{zr_1\varepsilon_{11}}{2RT}(1-\theta_{1(m)})^2 + r_1\left(\dfrac{1}{r_2} - 1\right) \quad (12)$$

This is equation 79 of Martire and Boehm's treatise (4). Thus it is shown that the apparently disparate models of Sanchez-Lacombe and Martire-Boehm both produce the same isotherm equation if it is assumed that $\tilde{\rho}_s = 1$ and thus do not need to be treated as distinct, independent models.

Equation 12 can also be cast in terms of a more classical reduced temperature and density based on the critical constants using relations derived from the equation-of-state,

i.e., $T_{R(1)} = \dfrac{T}{T_{c(1)}} = \dfrac{-RT}{z\varepsilon_{11}}\left\{\dfrac{(1+\sqrt{r_1})^2}{r_1}\right\}$ and $\rho_{R(1)} = \dfrac{\rho}{\rho_{c(1)}} = \theta_{1(m)}(1+\sqrt{r_1})$. This results in a slightly different isotherm equation with the fluid (mobile) phase parameters expressed in terms of $T_{R(1)}$ and $\rho_{R(1)}$.

$$\ln\theta_{1(s)} + r_1\left\{\dfrac{1}{r_2} - \dfrac{1}{r_1}\right\}\theta_{1(s)} + r_1\chi_{12}(1-\theta_{1(s)})^2$$
$$= \ln\left(\dfrac{\rho_{R(1)}}{(1+\sqrt{r_1})}\right) + r_1\left(1 - \dfrac{1}{r_1}\right)\left(\dfrac{\rho_{R(1)}}{(1+\sqrt{r_1})}\right) + \dfrac{(1+\sqrt{r_1})^2}{2T_R}\left(1 - \left(\dfrac{\rho_{R(1)}}{(1+\sqrt{r_1})}\right)\right)^2 + r_1\left(\dfrac{1}{r_2} - 1\right) \quad (13)$$

A version of this isotherm equation 13 has been applied previously (34, 35) for the CO_2-poly(methylmethacrolyate) systems and more detailed discussions of the assumptions and conditions used in the derivation are discussed therein and in the original publication (4). It was found for this system that the interaction parameter varied linearly with 1/T with little or no composition dependence.

Martire and Boehm's critical constant-based isotherm, equation 13, was fit to the isotherm data for CO_2 in PDMS in the same manner, i.e., with adjustable χ_{12} values, using the critical constants for CO_2 ($\rho_c = 0.468$ g/mL and $T_c = 31$ °C), the density of pure CO_2 calculated from the Ely equation-of-state, and an r value of 6.0. The results were very similar to those obtained for the Sanchez-Lacombe model, equation 11. The same composition dependence of the interaction parameter was observed whether the reduced temperature and density were calculated from the characteristic constants or the critical constants.

Conclusions

The independently derived models proposed by Sanchez-Lacombe and Martire-Boehm produce the same isotherm equation when the chemical potentials of CO_2 in the mobile and stationary phases are equated and the explicit pressure dependence is removed from the equations by use of an equation-of-state. The common model does not, however, adequately describe the experimentally observed solubility of CO_2 in PDMS without introducing a composition-dependent interaction term for the CO_2-polymer mixture.

Tracer pulse chromatography provides a simple, but very accurate, experimental method for the determination of the solubility or adsorbability of supercritical fluids, such as CO_2, in or on common SFC stationary phases. The method can be used for both pressures and temperatures encompassing the critical point. At pressures greater than $P_c =$ 74 atm, the experimentally measured solubility of CO_2 varies with experimental procedure. The chromatographic methods indicate only moderate solubility of CO_2 in PDMS at high pressures, whereas the more classical methods report much higher solubilities. One unique property of the very fast chromatographic methods is that the thermodynamic measurements are carried out with very thin films of polymers in order to achieve rapid equilibration of CO_2 distributed between the mobile and stationary phases. The gravimetric, volumetric, and ultrasonic experiments are usually performed on bulk polymer solutions to increase the sensitivity and accuracy of the solubility measurements. It is possible that the film thickness influences the solubility of CO_2 in polymers when the CO_2 is a supercritical fluid.

Acknowledgments

The research described herein was supported by the National Science Foundation. The GC/MS instrumentation was donated by the Hewlett Packard Company.

Literature Cited

1. Chester, T. L. *Anal. Chem.* **1997**, 165A-169A.
2. Giddings, J. C. *Unified Separation Science*: John Wiley & Sons, Inc, 1991.
3. Martire, D. E.; and Boehm, R. E. *J. Phys. Chem.* **1983**, *87*, 1045-1062.
4. Martire, D. E.; and Boehm, R. E. *J. Phys. Chem.* **1987**, *91*, 2433-2446.
5. Martire, D. E. *Liq. Chromatogr.* **1987**, *10*, 1569-1588.
6. Martire, D. E. *Liq. Chromatogr.* **1988**, *11*, 1779-1807.
7. Martire, D. E. *J. Chromatogr.* **1988**, *452*, 17-30.
8. Roth, M. *J. Phys. Chem.* **1990**, *94*, 4309-4314.

9. Roth, M. *J. Microcol. Sep.* **1991**, *3*, 173-184.
10. Roth, M. *J. Phys. Chem.* **1992**, *96*, 8548-8552.
11. Roth, M. *Supercritical Fluids* **1994**, 631-639.
12. Roth, M. *J. Chromatogr. A* **1996**, *738*, 101-114.
13. Roth, M. *J. Phys. Chem.* **1996**, *100*, 2372-2375.
14. Fleming, G.; and Koros, W. *Macromolecules* **1986**, *19*, 2285-2291.
15. Briscoe, B. J.; and Zakaria, S. *J. Polym. Sci. Polym. Phys. Ed.* **1991**, *29*, 989-999.
16. Garg, A.; Gulari, E.; and Manke, C. *Macromolecules* **1994**, *27*, 5643-5653.
17. Vincent, M. F.; Kazarian, S. G.; West, B. L.; Berkner, J. A.; Bright, F. V.; Liotta, C. L.; and Eckert, C. A. *J. Phys. Chem. B* **1998**, *102*, 2176-2186.
18. Shim, J.-J.; and Johnston, K. P. *A.I.Ch.E., J.* **1989**, *35*, 1097-1106.
19. Strubinger, J. R.; Song, H.; and Parcher, J. F. *Anal. Chem.* **1991**, *63*, 98-103.
20. Pope, D. S.; Sanchez, I. C.; Koros, W. J.; and Fleming, G. K. *Macromolecules* **1991**, *24*, 1779-1783.
21. Panda, S.; Bu, Q.; Huang, B.; Edwards, R. R.; Liao, Q.; Yun, K. S.; Parcher, J. F. *Anal. Chem.* **1997**, *69*, 2485-2495.
22. Edwards, R. R.; Tao, Y.; Xu, S.; Wells, P. S.; Yun, K. S.; and Parcher, J. F. *J. Phys. Chem. B* **1998**, *102*, 1287-1295.
23. Ely, J. F. *CO2PAC: A Computer Program to Calculate Physical Properties of Pure CO2*: Nat. Bur. of Standards, Boulder, CO, 1986.
24. Shih, H.; and Flory .P. J. *Macromolecules* **1972**, *5*, 759.
25. Flory, P. J. *J. Chem. Phys.* **1942**, *10*, 51.
26. Huggins, M. L. *Ann. N. Y. Acad. Sci* **1942**, *43*,
27. Sanchez, I. C.; and Lacombe, R. H. *J. Phys. Chem.* **1976**, *80*, 2352-2362.
28. Lacombe, R. H.; and Sanchez, I. C. *J. Phys. Chem.* **1976**, *80*, 2568-2580.
29. Sanchez, I. C.; and Lacombe, R. H. *Macromolecules* **1978**, *11*, 1145-1156.
30. Sanchez, I. C.; and Panayiotou, C. G., *Models for Thermodynamic and Phase Equilibrium Calculations*, S. Sandler, Ed., Marcel Dekker, New York, 1994.
31. Ely, J. F. *Proc. 63rd Gas Processors Assn. Conv.*, 1984.
32. Sanchez, I. C. *Macromolecules* **1991**, *24*, 908-916.
33. Condo, P. D.; Sanchez, I. C.; Panayiotou, C. G.; Johnston, K. P. *Macromolecules* **1992**, *25*, 6119-6127.
34. Parcher, J. F.; Edwards, R.R.; Tao, Y.; Xu, S.; Wells, S. P. and Yun, K.S. *J. Polymer Science Part B* **1998**, *36*, 2537-2549.
35. Parcher, J. F. *Chromatographia* **1998**, *47*, 570-574.

Techniques and Specific Applications

Chapter 8

Packed Capillary Columns in Hot Liquid and in Supercritical Mobile Phases

T. Greibrokk, E. Lundanes, R. Trones, P. Molander, L. Roed,
I. L. Skuland, T. Andersen, I. Bruheim, and B. Jachwitz

Department of Chemistry, University of Oslo, P.O. Box 1033, Blindern,
0315 Oslo, Norway

The advantages of packed capillary columns in liquid chromatography (HPLC) as well as in supercritical fluids have been demonstrated. In liquids the active use of temperature as a variable has resulted in reduced analyses time, increased column efficiency, increased signal to noise ratio, and allowed temperature programming for controlling retention and replacing solvent gradients in HPLC. Temperature programs have been shown to be compatible with detection by light scattering and inductively coupled plasma-mass spectrometry. Separation of glycerides could be obtained either by a pressure program in supercritical fluid chromatography (SFC) or by a temperature program in HPLC. For high temperature purposes, at pressures which keep solvents in the liquid state above their boiling points, compounds with low solubility at room temperature have been injected by a high temperature injector, separated with temperature programs to 160 °C and detected by evaporative light scattering. New synthetic organometallic catalysts which require more inert conditions have been purity tested in CO_2 by SFC on the packed capillaries and flame ionization detection. Large volume injections on the packed capillaries with 0.32mm i.d. have been demonstrated both in SFC (75 µL) and in HPLC (100 µL).

From the history of gas chromatography the virtues of (open tubular) capillary columns are well known and commonly accepted. Due to the increased viscosity and reduced diffusivity in supercritical fluids and particularly in liquids, the advantages of the open tubular columns become less prominent. At the same time a major disadvantage, reduced loadability, becomes more significant, primarily in the liquid state which requires very narrow columns in order to maintain high efficiency. Consequently, packed capillaries are a compromise which maintains a low thermal mass, allowing rapid temperature programming, while at the same time accepting larger injection volumes without overloading, and finally making available the wide selectivity range of column packings in liquid chromatography (HPLC). Packed columns cannot compete with open tubulars on the potential number of theoretical plates, due to the restrictions on column length, but in fluids of reduced viscosity high speed analyses can be performed on relatively long columns. Lowering the viscosity can be obtained by raising the temperature or by including fluids such as CO_2 in the mobile phase. In conventional HPLC, at room temperature, the back pressure allows only relatively short columns to be used. Narrow column diameters were early on advocated for HPLC (1), partly for environmental reasons, saving organic solvents, and partly for the increased signal to noise ratios with limited sample sizes. Although the environmental concerns are likely to increase in years to come, the ability to use the temperature actively for controlling retention is in our view another significant advantage with packed capillaries. Since HPLC started as a complimentary technique to GC, replacing temperature with solvent strength, the need for temperature control has appeared to be less urgent. Fears of decomposing analytes and stationary phases, sometimes unwarranted, have added to the negligence of temperature as a variable in liquids. The elution strength of aqueous reversed phase systems has been compared to variation of temperature (2,3). Thus, increasing temperature with 4-5 centigrades compares to increasing the methanol or acetonitrile concentration with 1% in a reversed phase system. Temperature programming in narrow columns in HPLC has been demonstrated on packed capillaries (2,4-7) and also on open tubular columns (8), but so far has not been brought into common use, partly due to reported problems with silicabased packings in aqueous phases.

In this paper we intend to demonstrate some of the advantages of packed capillary columns in liquid chromatography and in supercritical fluid chromatography (SFC). In liquids the temperature has been used actively for controlling retention and for replacing gradient elution, for increasing efficiency and lowering column back pressure at higher temperatures and for obtaining enhanced compatibility of some detectors with the low flow rates of such columns. Supercritical conditions have been utilized when the inertness of CO_2, flame ionization detection and the improved performance of modern column packings have been required.

Viscosity and Column Back Pressure

The viscosity of a liquid is given by:

$$\eta = A \exp(E_\eta / RT)$$

(E_η = effective activation energy for molecular displacement)

Since the viscosity is reduced with approximately 1-2 % per centigrade (9), increasing the temperature from 20 to 50 °C can result in a significant reduction of the viscosity and consequently in the column back pressure:

$$P = \eta u L / d_p^2$$

The 20-80 cm long columns, with i.d. of 100-320 μm, were packed (7) with 3-5 μm particles. Typical flow rates for the 0.32 mm columns in HPLC were 5 μL/min. The reduced viscosity can be utilized for increasing the column length (L), if needed, reducing the particle size (d_p) or simply for increasing speed (u) and reducing the time of analysis. This is illustrated in the determination of retinoids in Figure 1, where a 100 μL volume of the sample solution was injected on to a packed capillary column. In order to avoid overloading the column, the elution strength of the sample solvent was lower than the mobile phase, allowing analyte focusing at the column inlet. The sample loading time of a 100 µl sample was reduced from 20 min at room temperature to 5 min at 50 °C.

Temperature - Column Efficiency

Increasing the temperature will usually be expected to increase the column efficiency, due to the increased diffusivity which reduces the C-term of the van Deemter equation. Thus, by increasing the temperature from 25 to 70 °C, the reduced plate height of all-trans retinoic acid was halved, as shown in Figure 2. As a consequence of the reduced retention and the increased efficiency, the peak height increased by a factor of 2.8, while the baseline noise remained the same.

Determination of retinoids in blood samples, particularly from children (small samples), is an example of applications where improved signal to noise ratio of capillary columns vs. conventional HPLC columns can be utilized. A column temperature of 50 °C appeared to have no adverse effects neither on the retinoids nor on the column packing.

In capillary electrochromatography on packed capillaries, where the mechanical pump is replaced by an electrical field, the relationship between column efficiency and temperature is not as simple (Figure 3). By increasing temperature from 20 to 60 °C, the column efficiency decreased in a non-aqueous system, but the efficiency per time unit increased as the EOF increased by 32% (Roed, L.; Lundanes, E.; Greibrokk, T., *J. Microcol. Sep.*, in press).

Temperature - Retention in HPLC

The effect of temperature on retention in aqueous systems has been referred to previously. In non-aqueous systems, solvent gradients are of little or no use either in adsorption chromatography (10) or in gel permeation chromatography. In adsorption chromatography this is a result of the lack of reproducibility of gradient elution due to varying activity of the surface functions. If differences in adsorption energies are small, van´t Hoff plots will be similar to Figure 4, resulting in proportional reduction of retention by increasing temperature. With larger differences in adsorption energies, temperature programs may implement changes comparable to selectivity changes of a solvent gradient.

In non-aqueous systems, which usually makes use of isocratic or stepwise elution, a continuous retention control can be obtained by temperature programming on packed capillaries (11). Table 1 shows the effect on retention of a temperature increase from 50 to 150 °C.

At temperatures exceeding the boiling points of the mobile phase components, a restrictor has been connected to the outlet of the system, maintaining a back pressure which prevents the solvents from boiling inside the column. Figure 5 compares isothermal elution of polymer additives at 50 °C with a temperature program from 50 to 150 °C. Retention times were reduced and peak shapes and detection limits were improved by the temperature programming. The precision of the retention times measured during this temperature program was determined to 3% RSD (n=6).

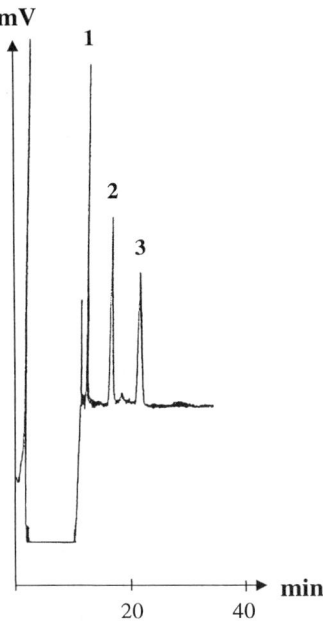

Figure 1. On-column focusing large volume (100 µL) injection of a 50 ppb solution of all-*trans* retinol (1), 13-*cis*-retinoic acid (2) and all-*trans* retinoic acid (3) on a 0.32 x 250 mm fused silica column, packed in house with 5 µm Suplex pKb-100. Sample introduction flow; 20 µL/min, operating temperature; 50 °C, sample solvent; acetonitrile-0.5% ammonium acetate in water-water (45:5:50), mobile phase; acetonitrile-0.5% ammonium acetate in water-acetic acid (94.9:5.0:0.075), mobile phase flow; 5 µL/min, UV-detection at 360 nm. (Adapted from Molander, P.; Gundersen, T.E.; Haas, K.; Greibrokk, T.; Blomhoff, R.; Lundanes, E., *J. Chromatogr. A,* in press)

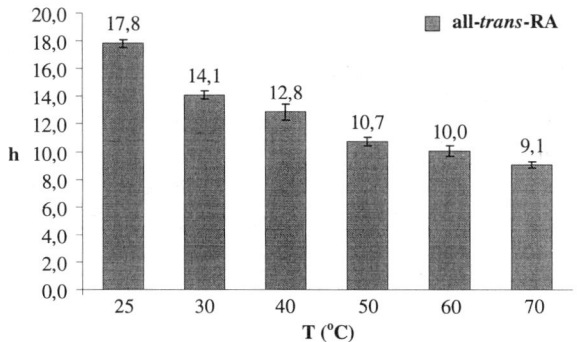

Figure 2. Effect of temperature on the reduced plate height (h) of all-*trans*-retinoic acid in the liquid chromatographic system described in Figure 1. (Adapted from Molander, P.; Gundersen, T.E.; Haas, K.; Greibrokk, T.; Blomhoff, R.; Lundanes, E., *J. Chromatogr. A,* in press)

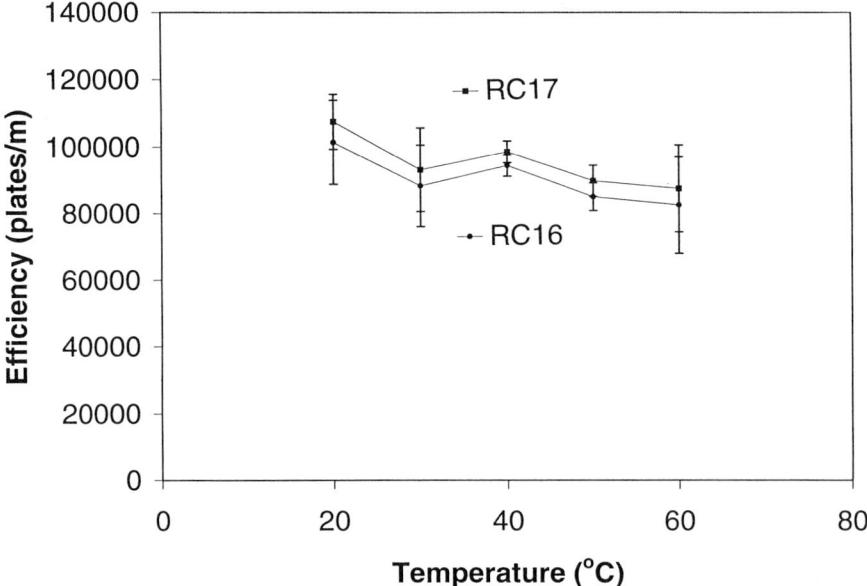

Figure 3. Column efficiency (plates/m) of all-*trans*-retinyl hexadecanoate (RC16) and -heptadecanoate (RC17) as a function of column temperature in electrochromatography with 2.5 mM lithium acetate in dimethyl formamide-methanol (99:1) on 3 µm Hypersil ODS in a 0.18 x 300 mm fused silica column at 650 V/cm. (Reproduced with permission from Roed, L.; Lundanes, E.; Greibrokk, T., *J. Microcol. Sep.*, in press. Copyright 1999 John Wiley & Sons.)

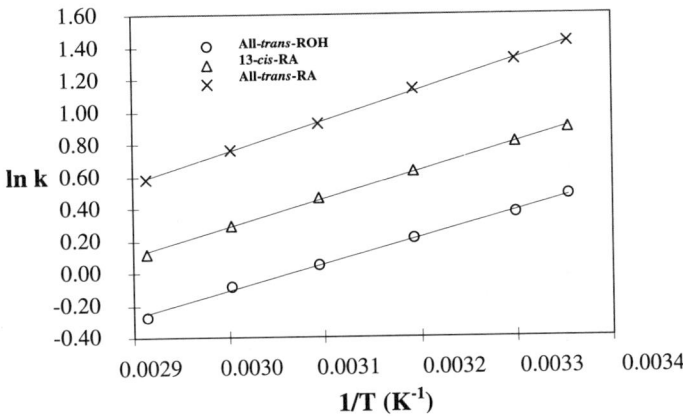

Figure 4. Van´t Hoff plot of all-*trans* retinol, 13-*cis* retinoic acid and all-*trans* retinoic acid. (Adapted from Molander, P.; Gundersen, T.E.; Haas, K.; Greibrokk, T.; Blomhoff, R.; Lundanes, E., *J. Chromatogr. A,* in press.)

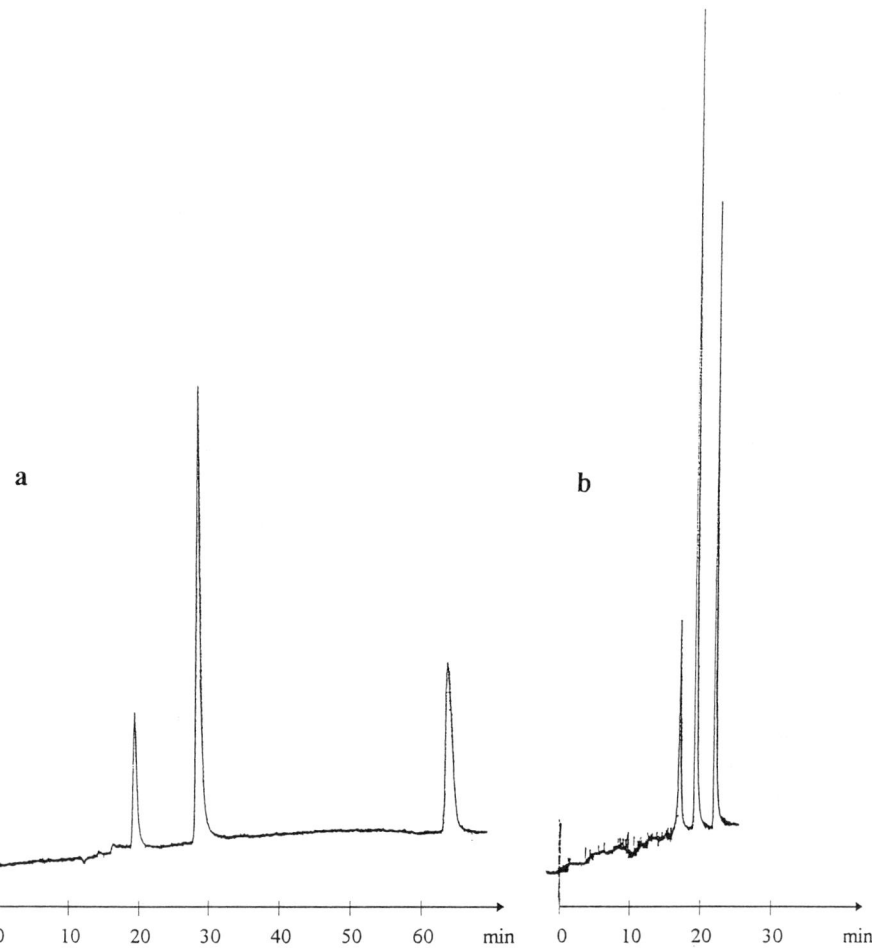

Figure 5. Comparing retention and peak shape of the polymer additives Irganox 3114, Tinuvin 327 and Irganox 1076 at constant temperature of 50 °C (a) and a temperature program to 150 °C (b). Chromatographic conditions as in Table I.

Table I. Retention factor (k) of octadecyl-3-(3,5-di-t.butyl-4-hydroxyphenyl)-propionate as a function of temperature (T). Mobile phase: acetonitrile - dimethylformamide (90:10). Column: 0.32 x 700 mm fused silica packed with 5 μm Hypersil ODS.

$T\ (°C)$	k
50	2.86
75	1.31
100	0.70
125	0.38
150	0.24

Detector Compatibility and Applications with Packed Capillaries

One of the major advantages of SFC is the ability to use GC detectors, with CO_2 as mobile phase. Glycerides can be separated by SFC as well as by high temperature liquid chromatography (HTLC). In plain CO_2 on an open capillary column the glycerides were separated by a pressure program and detected by FID (Figure 6). A packed column required a modifier to obtain good peak shapes. In the latter case as well as with an HTLC separation in 10% ethyl acetate in acetonitrile on a packed capillary, an evaporative light scattering detector (ELSD) replaced the FID. Similar resolution was obtained with the temperature program in HTLC as with the density program in SFC. The temperature program from 40 to 160 °C had no effect on the baseline. In a more high-boiling solvent and with high ramp rates baseline changes have appeared (12). With the low flow rates (ca 5 μL/min) the peak area is almost linearly related to the mass injected (12), in contrast to the common perception of the light scattering detector as being a non-linear detector.

Another area where supercritical fluid chromatography has demonstrated high usefulness is in the determination of polymer additives. Early work showed that some of the additives could be chromatographed in plain CO_2, while others needed modifiers, then excluding the use of the FID. With the steadily improved quality of column packings, more additives can now be eluted with good peak shapes in CO_2, with FID, as demonstrated in Figure 7. Not all additives can be chromatographed by SFC, though, due to insufficient solubility in CO_2. A typical example is Chimasorb 944, a complex polymer additive with MW up to and above 2500. The best resolution ever of this very difficult mixture was obtained with high temperature LC and a temperature program to 160 °C (Figure 8).

At the low flow rates of the packed capillaries the FID is also compatible with plain water as mobile phase (Figure 9). At room temperature the three butanols shown were strongly retained on the Hypercarb column, while a temperature increase to 50°C resulted in reduced retention and greatly improved peak shapes.

Another detection method which has been studied was inductively coupled plasma-mass spectrometry, where the original nebulizer wasting 95-99% of the

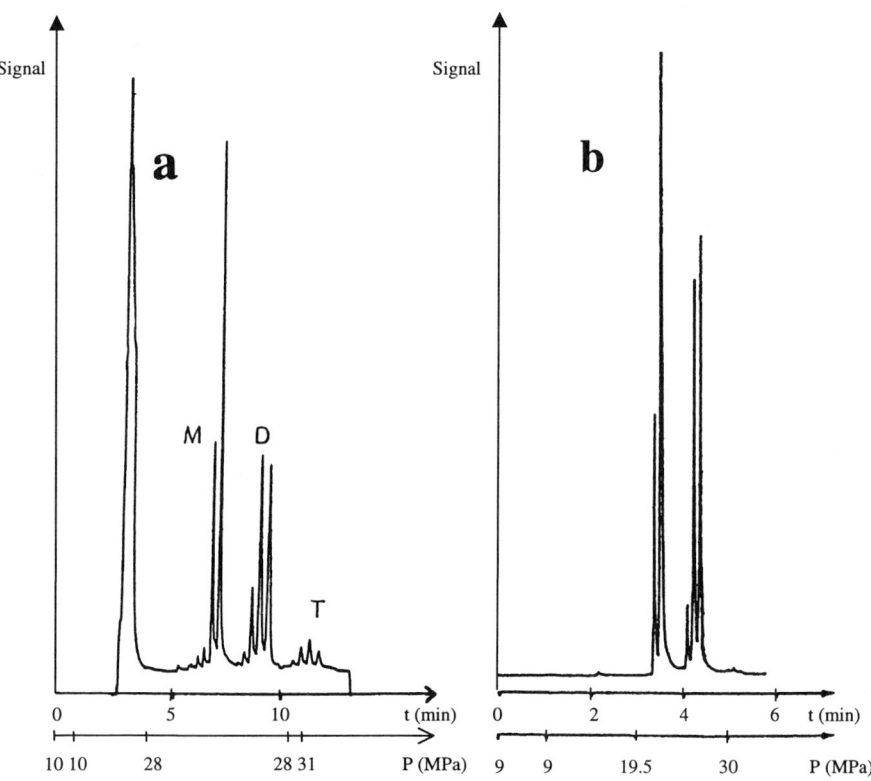

Figure 6. Separation of monoglycerides (M), diglycerides (D) and triglycerides (T) in a technical product by open tubular SFC on 50 μm x 10 m SB-Biphenyl-30 with pressure program in CO_2 with FID (a), packed capillary SFC on 0.32 x 100 mm fused silica with 4 μm Novapak C18 with pressure program in 2.9 mol% 1-propanol in CO_2 with ELSD (b) and packed capillary HTLC on 0.32 x 700 mm fused silica with 5 μm Hypersil BDS with temperature program from 40 to 160 °C in acetonitrile-ethyl acetate (90:10) with ELSD (c).

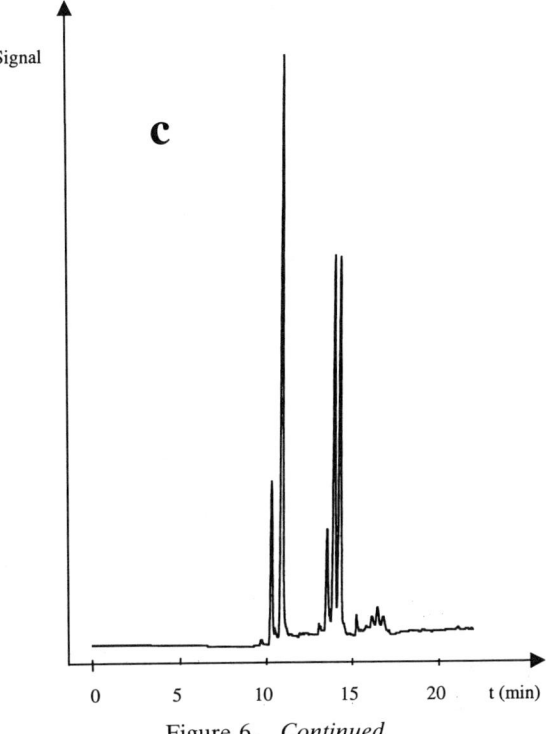

Figure 6. *Continued*

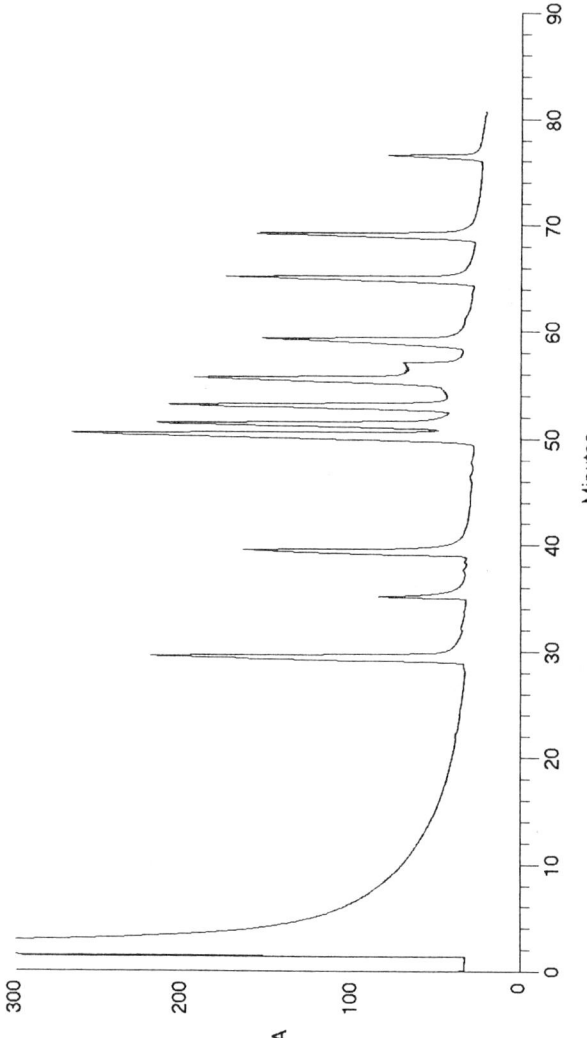

Figure 7. Separation of polymer additives by packed capillary column SFC in plain CO_2 on 0.32 x 360 mm fused silica column with 5 μm Kromasil C18 at 100 °C with a pressure program from 125 to 380 bar. The peaks are in order of elution; n-eicosane (internal standard), glycerides (next two peaks), α-Tocopherol, Irgafos 168, Irgafos 168-phosphate, Irganox 1076, Irganox 1035, Irganox 3114, Irganox 1330, Irganox MD 1024 and Irganox 1010.

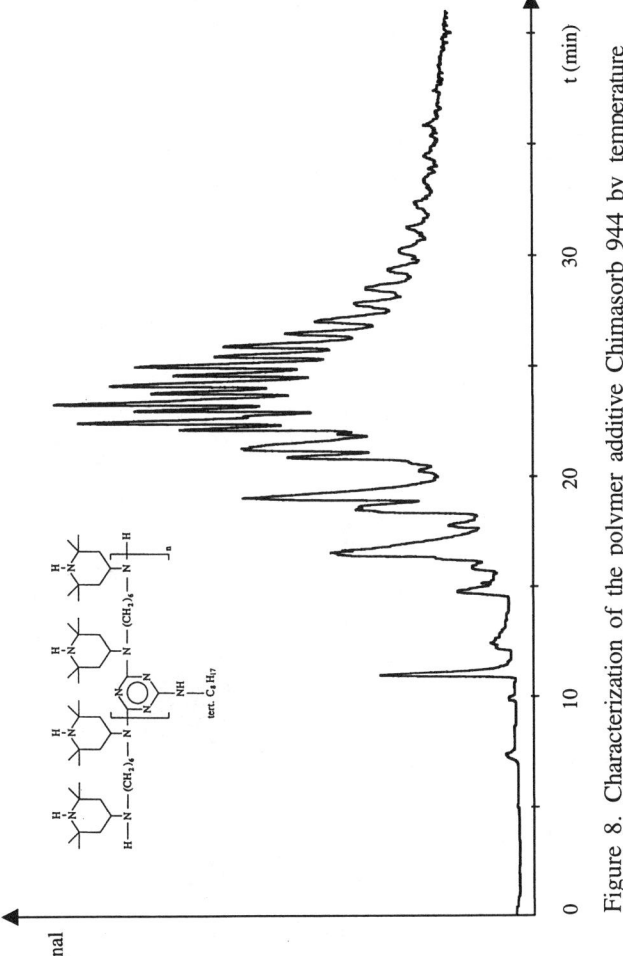

Figure 8. Characterization of the polymer additive Chimasorb 944 by temperature programming from 80°C (15 min), then 20°C/min to 160°C, in ethyl acetate-methanol-acetonitrile (45:5:50) on 0.32 x 700 mm fused silica with 5 μm Hypersil BDS and evaporative light scattering detection.

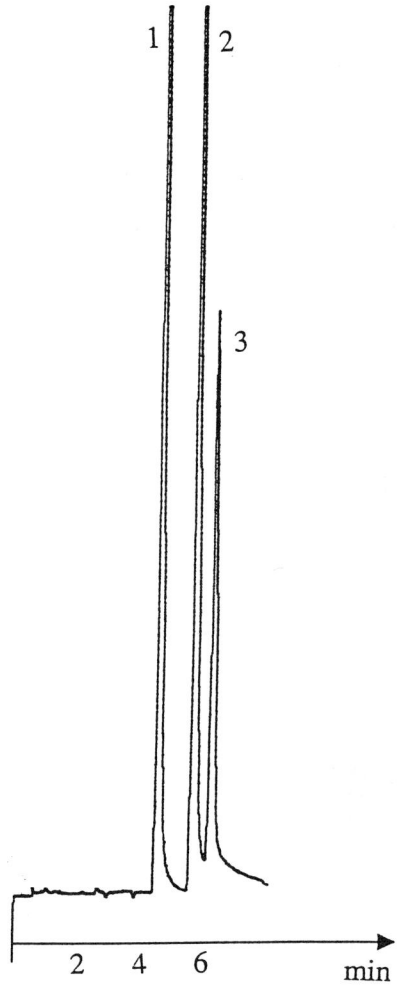

Figure 9. Flame ionization detection in HPLC in water (5 µL/min) of t.butanol (1), sec. butanol (2) and n-butanol (3) after separation on 0.25 x 250 mm fused silica with 5 µm Hypercarb at 50 °C. Detector temperature; 350 °C, with 35 mL/min of H_2 and 240 mL/min of air.

sample was replaced by a non-splitting direct inlet nebulizer (13). A temperature program had no effect on the baseline (Figure 10).

In capillary electrochromatography UV-detection imply sensitivity limitations due to the short light path of on-column detection. As a compromise between efficiency and solubility in a non-aqueous electrochromatographic system, 40 °C was chosen for determination of retinyl palmitate in polar fox liver (Figure 11). The low currents in this mobile phase permitted the use of larger inner diameter columns (180 µm) than normal, without excessive Joule heating, resulting in 6-7 times higher peak heights than with 100 µm i.d. columns (Roed, L.; Lundanes, E.; Greibrokk, T., *J. Microcol. Sep.*, in press).

Organometallic catalysts for single site polymerization reactions are frequently rather unstable compounds, reactive towards humidity and oxygen. Purity testing of new organometallic compounds have been performed in CO_2 with detection by FID In order to prevent contact with air, a miniaturized combination of extraction and chromatography in CO_2 was performed (Figure 12). Solid catalyst, synthesized in house, was extracted by CO_2, focused at the column inlet and analyzed by SFC, completely avoiding the use of solvents. Thus, the exposure towards the atmosphere was reduced and potential biproducts which otherwise would be covered by the solvent tail during analysis, could be detected.

Large volume injections on capillary columns are relatively easy in aqueous reversed phase HPLC, by injecting in solutions of low solvent strength. In SFC it is more difficult to find an organic solvent with clearly lower elution strength than CO_2. Thus, combining SFE and SFC may become the only alternative, as shown in Figure 13. A 75 µL extract of apples was applied on a small solid phase extraction column, the solvent removed and the pesticide extracted with SFE and transferred to and focused on the packed capillary column.

High Temperature Applications

High molecular weight waxes, resins and many polymers are not soluble in organic solvents at room temperature and need heated solvents to stay in solution. Size exclusion chromatography of polyolefins are for example routinely performed at 140 °C. Many hydrocarbon waxes precipitate during injection in ordinary HPLC injectors. Since there are no heated injectors which are compatible with the capillary columns in HPLC commercially available, a heated injector has been constructed (Figure 14). With this injector hydrocarbon waxes were injected and separated with temperature programs (Figure 15). The authors intend to utilize this injector in improving methods for characterization of polymer alloys and mixtures, combined with temperature programming on packed capillaries, for separations according to functionality as well as size.

Robustness of Columns

Packed fused silica columns need to be handled more carefully than conventional steel HPLC columns, but are still fairly tolerant towards normal handling. Dirty samples should be filtered to avoid plugging of the inlet frit, as with conventional columns. Well packed columns with silica based packings have tolerated more than 400 temperature programs to 150 °C in non-aqueous solvents without significant changes in retention and efficiency. In aqueous solvents silica based packings have previously been known for not withstanding temperatures above 70-80 °C without

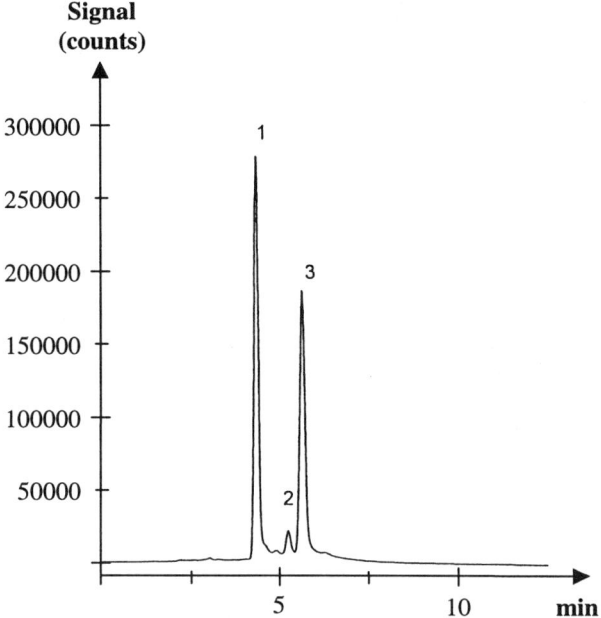

Figure 10. Temperature programming from 50°C, then 16 °C/min to 100°C on 0.32 x 230 mm fused silica with 5 µm Kromasil C18 in acetonitrile with inductively coupled plasma mass spectrometric detection. Separation of tetramethyl lead (1) from tetraethyl lead (3) and an impurity (2). (Adapted from Trones, R: Tangen, A.; Lund, W.; Greibrokk,T., *J. Chromatogr. A.*, in press.)

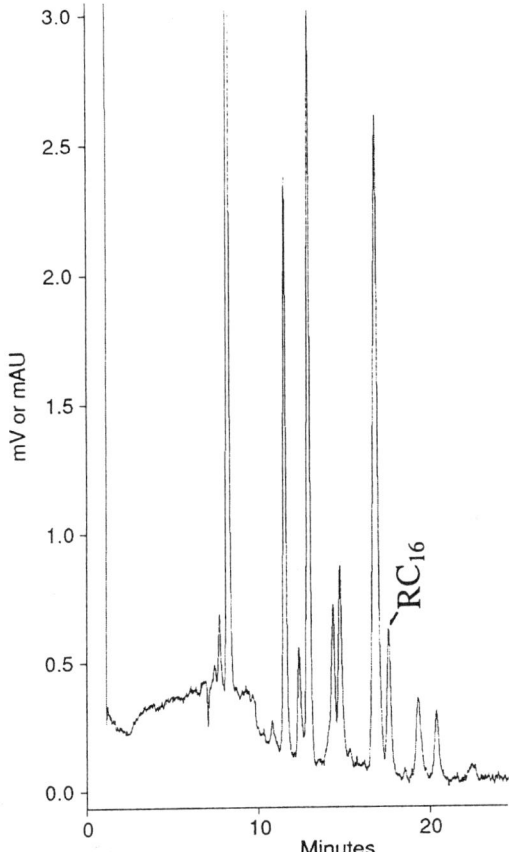

Figure 11. Retinyl ester profile of liver extracts from polar fox, including the determination of retinyl palmitate (RC16) by non-aqueous electrochromatography on 0.18 x 210 mm fused silica packed with 3 μm Hypersil ODS, in 2.5 mM lithium acetate in DMF-methanol (99:1) at 40 °C and 650 V/cm. (Reproduced with permission from Roed, L.; Lundanes, E.; Greibrokk, T., *J. Microcol. Sep.*, in press. Copyright 1999 John Wiley & Sons.)

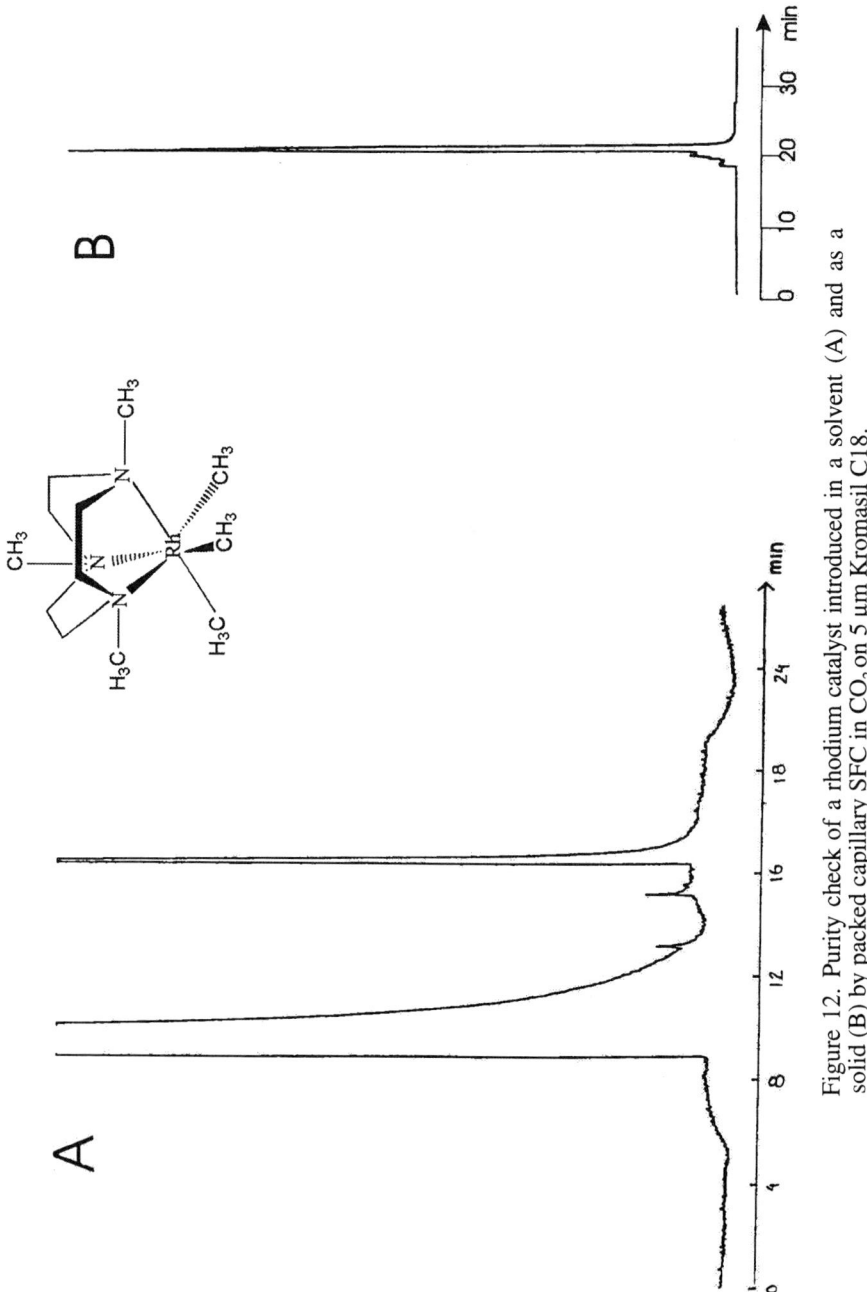

Figure 12. Purity check of a rhodium catalyst introduced in a solvent (A) and as a solid (B) by packed capillary SFC in CO_2 on 5 μm Kromasil C18.

137

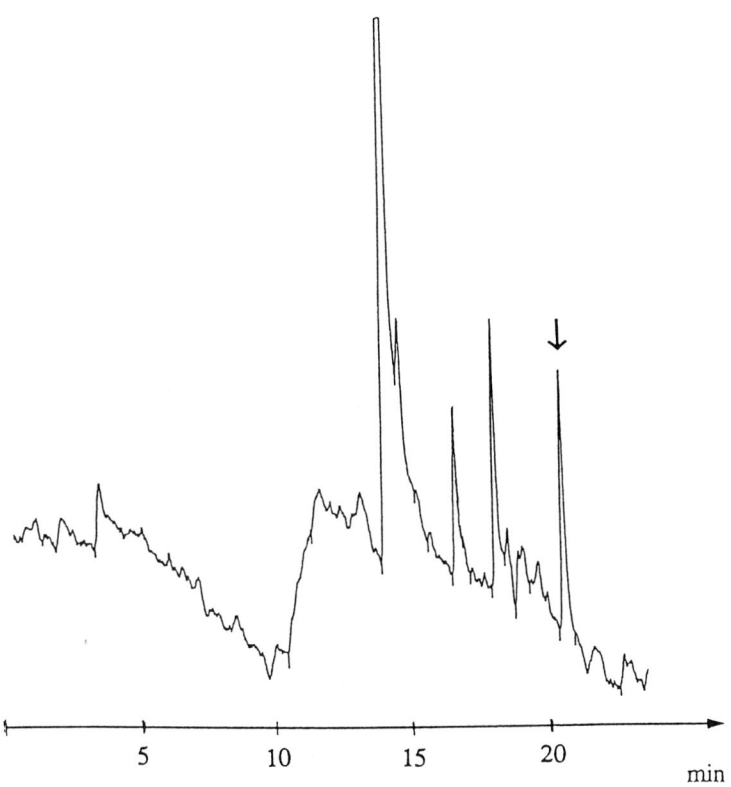

Figure 13. On-line combined solid phase injection-SFE-SFC in CO_2 of a 75 μL methanol/water apple extract containing the pesticide fenpyroximate (arrow). Column ; 0.32 x 530 mm fused silica with 5 μm Kromasil C18, pressure program from 100 to 345 bar, column temperature; 70 °C.

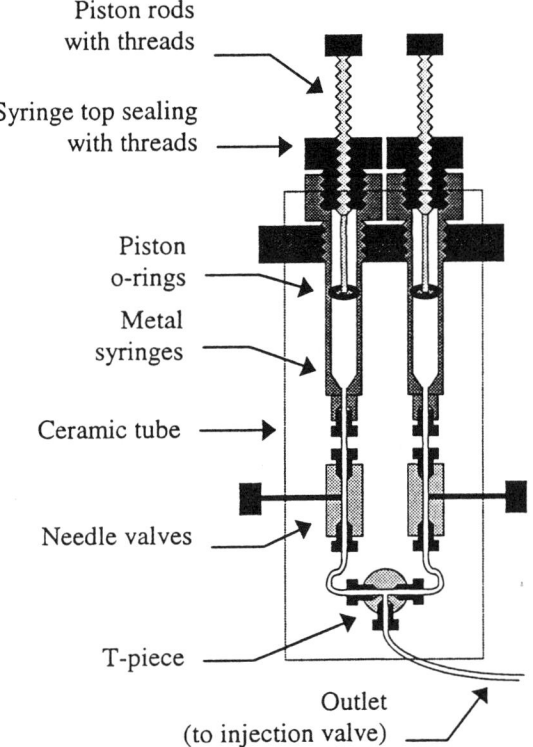

Figure 14. High temperature injector in HTLC.

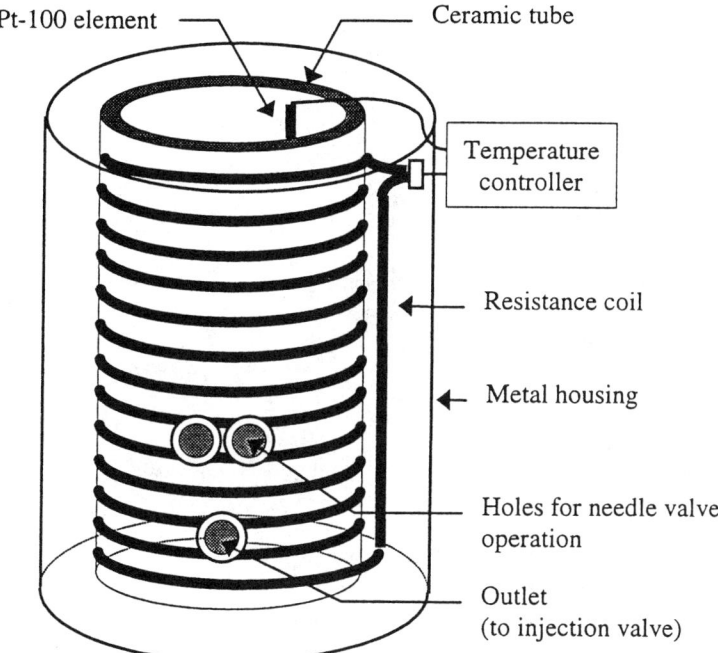

Figure 14. *Continued*

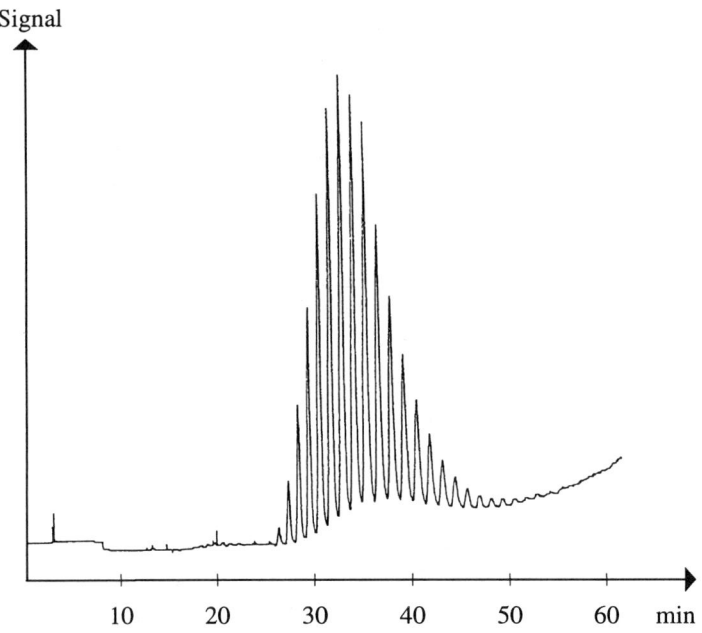

Figure 15. Injection of hydrocarbon wax (C14-C80) in cyclohexane at 70 °C. Separation by HTLC in acetonitrile - methylisobutyl ketone (75:25) on 0.32 x 1000 mm fused silica with 5 μm Kromasil C18 and light scattering detection. Temperature program starting at 70 °C for 10 min, then 1 °C/min to 130 °C.

being degraded, depending also on the pH. More recently made packings need to be tested to see whether this still is the case. As better silica qualities and improved coating techniques have allowed wider pH ranges of many reversed phase packings, an improved resistance towards higher temperatures in aqueous solutions might very well be expected. In aqueous solutions, temperatures higher than 70-80 °C will probably not be required for controlling retention and even temperatures of 50-60 °C may be sufficient for replacing gradient elution. Higher temperatures will be expected to have an extra effect on tailing peaks caused by non-intentional adsorption, due to the reduction in adsorption energy at higher temperature.

Thus, as soon as reversed phase columns can be demonstrated to withstand such temperatures over extended periods of time, at various pHs, temperature programming will be able of challenging gradient elution, as a much simpler and faster technique with packed capillary columns. In non-aqueous mobile phases, the advantages have already been proven.

Literature Cited

1. Takeuchi, T.; Watanabe,Y.; Ishii, D. *J. High Resolut. Chromatogr.* **1981**, *4,* 300
2. Bowermaster, J.; Mc Nair, H. *J. Chromatogr.* **1983**, *279* , 431
3. Vailaya, A.; Horváth, C. *J. Phys. Chem. B* **1997**, *101*, 5875
4. Hirata, Y.; Sumiya, E. *J. Chromatogr.* **1983**, *267*, 125
5. Jinno, K.; Phillips, J.B.; Carney, D.P. *Anal. Chem.* **1985**, *57* , 574
6. Yoo, J.S.; Watson, J.T.; McGuffin, V.L. *J. Microcol. Sep.* **1992**, *4*, 34
7. Trones, R.; Iveland,A.; Greibrokk, T. *J. Microcol. Sep.* **1995**, *7*, 505
8. Liu, G.; Djordjevic, N.M.; Erni, F. *J. Chromatogr.* **1992**, *592*, 239
9. Giddings, J.C. *Unified Separation Science*, John Wiley & Sons, New York, **1991**, 80
10. Snyder, L.R.; Kirkland, J.J. *Introduction to Modern Liquid Chromatography"*, John Wiley & Sons, New York, 1979, 391
11. Trones, R.; Iveland, A.; Greibrokk, T. *Proc. 6th. Int. Symp. on SFC/SFE*, Uppsala, Sept. 6-8, 1995
12. Trones, R.; Andersen, T; Hunnes, I.; Greibrokk, T. *J. Chromatography A* **1998**, *814*, 55
13. Tangen, A.; Trones, R.; Greibrokk, T.; Lund, W. *J. Anal. At. Spectrom.* **1997**, *12* , 667

Chapter 9

Packed Capillary Column Chromatography with Gas, Supercritical, and Liquid Mobile Phases

Keith D. Bartle[1], Anthony A. Clifford[1], Peter Myers[1], Mark M. Robson[3], Katherine Seale[1], Daixin Tong[1], David N. Batchelder[2], and Suzanne Cooper[2]

[1]School of Chemistry and to [2]Department of Physics and Astronomy, University of Leeds, Leeds S2 9JT, United Kingdom

Introduction

The concept of unified chromatography was defined by Giddings over thirty years ago, when he pointed out (1) that there are no distinctions between chromatographic separation modes which are merely classified according to the physical state of the mobile phase (GC, SFC and HPLC). Recently, Chester has described (2) how a consideration of the phase diagram of the mobile phase shows that a one-phase region (Figure 1) is available for the setting of mobile phase parameters, and that the boundaries separating individual techniques are totally arbitrary. By varying pressure, temperature and composition, solute - mobile phase interactions can be varied so as to permit the chromatographic elution of analytes ranging from permanent gases to ionic compounds; the dependence of solute diffusion coefficient in the mobile phase on pressure, temperature and composition, (Figure 1) also influences mass-transfer characteristics and also has an important bearing on the choice of an appropriate mobile phase.

Towards the end of the 1980s, the concept arose of using a single chromatographic system to carry out separation in different modes; the principles and applications of unified chromatography have recently been reviewed (3). The purpose of this paper is to show how: capillary columns with i.d. in the range 50 to 500 µm and packed with

[3]Current address: Express Separations Ltd., 175 Woodhouse Lane, Leeds LS2 3AR, United Kingdom.

(generally bonded) silica particles with supercritical CO_2 carrier permit GC, SFC and HPLC in the same chromatograph; (b) such columns can be used in capillary electrochromatography (CEC); and (c) that micro Raman spectroscopy is a promising detector for microchromatography.

Packed capillary columns offer the substantial advantages of small volumetric flow rates (1-20 µL min^{-1}) which have environmental advantages, as well as permitting the use of 'exotic' or expensive mobile phases. Peak volumes are reduced, driven by the necessity of analysing very small (picomole) amounts of substance available, for example, in small volumes of body fluids, or in the products of single-bead combinational chemistry.

Packing capillary columns

Capillary columns have most commonly been packed with <10 µm i.d. particles by liquid slurry methods, usually involving low viscosity solvents such as acetone; dry-packing methods have also found limited application for 250 µm i.d. columns, while centrifugal and electrokinetic procedures have also been proposed for the packing of columns for CEC. We have found that supercritical CO_2 is the most useful packing carrier for a full range of column internal diameters and particle diameters, allowing long (>1 m) columns to be packed, and packing medium density and viscosity to be varied by changing the applied pressure and temperature (4). Packing material is well dispersed in liquid CO_2 in a reservoir at room temperature, and packed under supercritical conditions (above 32°), maintained by a restrictor at the column exit which can be changed to vary the packing velocity. The column is subjected to ultrasonication during packing.

Columns packed in this way, and tested in SFC are HPLC are highly efficient and show the classical van Deemter behaviour (Figure 2) with the minimum in the reduced plate height (h´) shifted to faster carrier linear flow rates u and a flatter curve at high u because of a higher solute diffusion coefficient. An investigation of the dependence of column performance on packing variables revealed that: sonication is vital to avoid column voids; higher packing pressure gives better efficiency, with

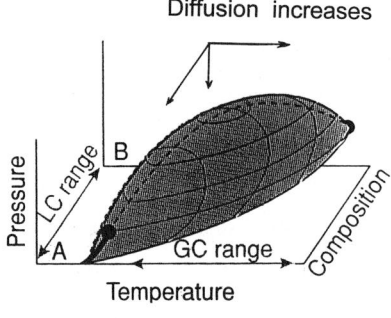

UNIFIED CHROMATOGRAPHY

Vary mobile phase to suit problem in single

Figure 1. Three dimensional two-component phase diagram. Shaded area is two-phase region. After reference 2.

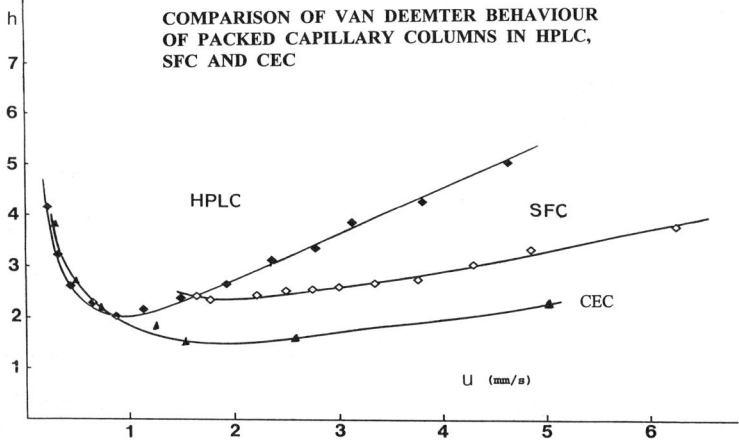

Figure 2. Van Deemter plots of reduced plate height, h, *versus* linear mobile phase velocity for packed capillary columns in HPLC, SFC and CEC. Columns: HPLC and SFC, 30 cm × 250 μm packed with Water Spherisorb ODS-2 5 μm; CEC 50 μm × 25 cm packed with Waters Spherisorb PAH 3 μm. Test solutes: HPLC, pyrene ($k' = 6.7$); SFC, chrysene ($k' = 5.4$); CEC, phenanthrene ($k' = 5.0$).

lower minimum reduced plate heights (h'_{min}) ~2; but, and counter-intuitively, looser, more porous packing results from higher packing pressures.

In fact, a comparison with literature h' values for columns packed by liquid slurry methods (5) show the same trend as our results (Figure 3). For dry-packed (6)

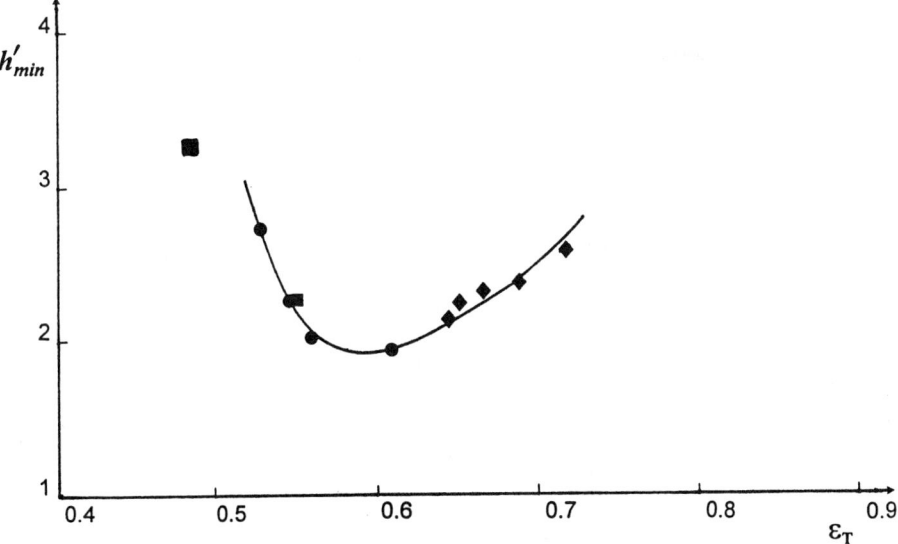

Figure 3. Plot of minimum reduced plate height, h_{min} versus total porosity (ε_T) for capillary columns packed with: (●) supercritical fluid carrier; (◆) dry-packing (reference 6); and (■) liquid slurry (reference 5).

columns there is an opposite trend (Figure 3), presumably because frictional behaviour is quite different during this procedure. The observation of looser packing with increasing pressure can be accounted for by a model (7) in which a greater slurry velocity during packed bed formation in the column centre produces a 'dome' of particles which rearrange laterally in a process which is favoured by reduced packing fluid density. The better efficiency of loose packed columns can be explained by a wall effect in which, for a tightly packed column bed, flow near the wall differs more from flow in the bed core than in a loosely packed column where there is a greater similarity of flows in the wall and core regions.

A unified chromatograph for GC, SFC and HPLC

The elements of the unified chromatograph (Figure 4) are: a helium cylinder with two-stage pressure regulator; syringe pump; reciprocating pump; injection valve with pneumatic actuator and digital valve sequence programmer; packed capillary column located in an oven; flame ionisation detector; and UV/visible detector with small volume flow cell (<100 nL). For SFC and HPLC either CO_2 from the syringe pump on liquid mobile phase supplied by the reciprocating pump are selected by positioning the mode selector valve. In GC mode, high pressure helium is supplied. Flow eluent from the column is directed either to the FID through a frit restrictor, or directly to the UV detector. If CO_2 is the mobile phase, a restrictor is located after the UV detector and used to control flow rate by varying the temperature of the surrounding heater.

High pressure GC on packed capillary columns in the unified chromatograph

The unified chromatograph can be readily used for GC on capillary columns packed with HPLC stationary phases. The small column volumes mean that gas inlet pressures up to 400 bar can be safely employed. Giddings (9) first showed the advantages of high pressures and small particle diameters in GC, and used pressures up to 2000 bar to analyse polymers and biomolecules. In the work described here we have separated hydrocarbons by GC on 5 µm ODS-bonded silica particles prepared by a supercritical bonding process (10). Figure 5 shows a typical chromatogram. Selectivity could be introduced into the separations by different carbon loadings on

A unified chromatograph for GC, SFC, HPLC and sequential GC-SFC analyses

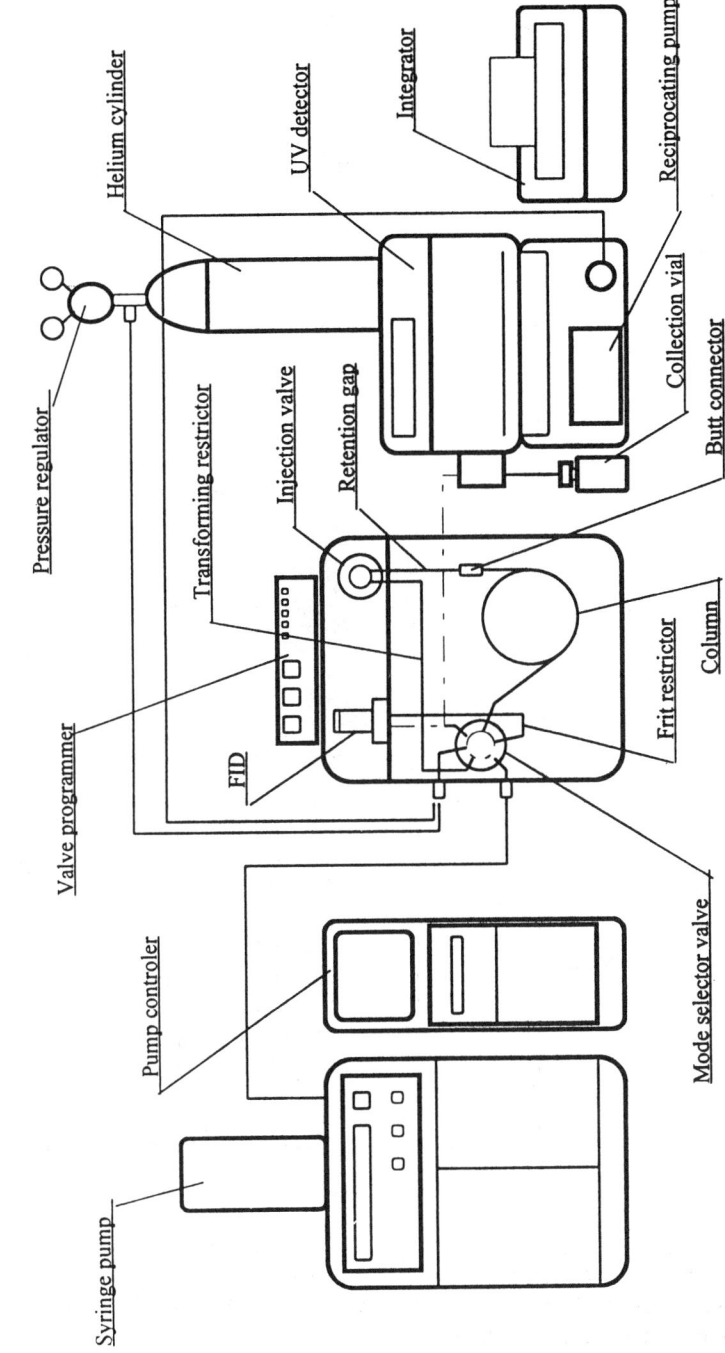

Figure 4. Schematic diagram of a unified chromatograph for GC, SFC and HPLC on packed capillary columns (after reference 8).

HPGC OF HYDROCARBONS

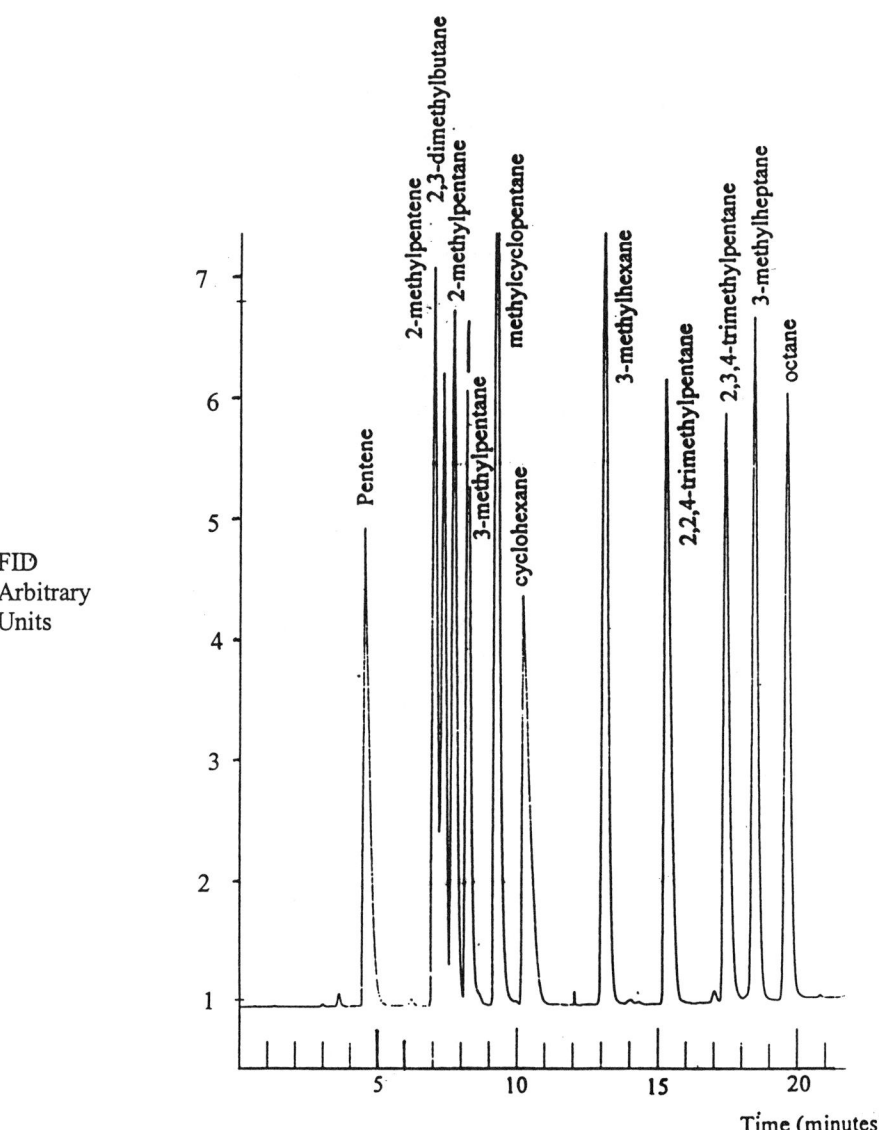

Figure 5. GC chromatogram of hydrocarbon standards on a packed capillary column (21.5 cm × 250 μm i.d. packed with 5 μm ODS-2) in the unified chromatograph. Mobile phase helium at 120 bar. Column temperature 40°C to 220°C at 5°C mm^{-1}. Detection by flame ionisation.

the particles; for example better resolution of the dimethylbutanes was achieved by reducing the carbon loading. Very low molecular weight (C_1-C_3) hydrocarbons could be separated at temperatures near ambient by high pressure GC on a capillary column packed with a carbon molecular sieve. This result opens up the possibility of the simple separation of isotopically labelled compounds (11) on carbon coated columns under accessible conditions. The GC van Deemter curves for ODS coated silica show very little increase in plate height after the minimum is reached with increasing mobile phase velocity. This shape of the curve is in agreement with the theory of Giddings (12) and the experimental work of Huber (13) and shows that the controlling factor is C_k, a mass transfer term describing the kinetics of adsorption and desorption from the solid surface. In fact, GC on ODS coated silica is best described as gas solid chromatography.

The separations described here differ from those by solvating gas chromatography discussed by Shen and Lee (14). The particular advantage of high pressure GC is the rate of generation of theoretical plates, compared in Table 1 with those achievable (15) by other chromatographic methods, and proportional to C_k^{-1}. In principle, very large numbers of plates per second are possible since C_k is very small.

SFC and HPLC on packed capillary columns in the unified chromatograph

The stability of supercritical CO_2 packed capillary columns for SFC and HPLC is very satisfactory. No voids in the columns appear over months of operation. Because of the compressibility of the supercritical mobile phase, rapid reductions of operating pressure are best avoided, but nonetheless, even sudden release of inlet pressure from 100 bar to ambient does not reduce column efficiency.

Examples of the use of packed capillary columns in the unified chromatograph for HPLC and SFC have made use of the wide range of silica-bonded stationary phases available either commercially or by in-house (10) bonding procedures. Figure 6 shows the high efficiency HPLC separation of nitrated polycyclic aromatic hydrocarbons on a 40 cm long ODS column. Simulated distillation of high molecular weight petroleum derivatives was achieved at low temperature by SFC with CO_2 on alkylsilyl bonded silicas. The closest correspondence between retention and boiling point for aromatics, alkylaromatics and alkanes was obtained on C_6-bonded silica

Table 1: Summary of plates per second achievable by different chromatographic techniques

Technique	Plates s^{-1}
LC packed	14-35
SFC packed	31-83
SFC open tubular	11-33
GC open tubular	93-180
High pressure GC (this work$^+$)	130
$^+$ other data taken from ref.15	

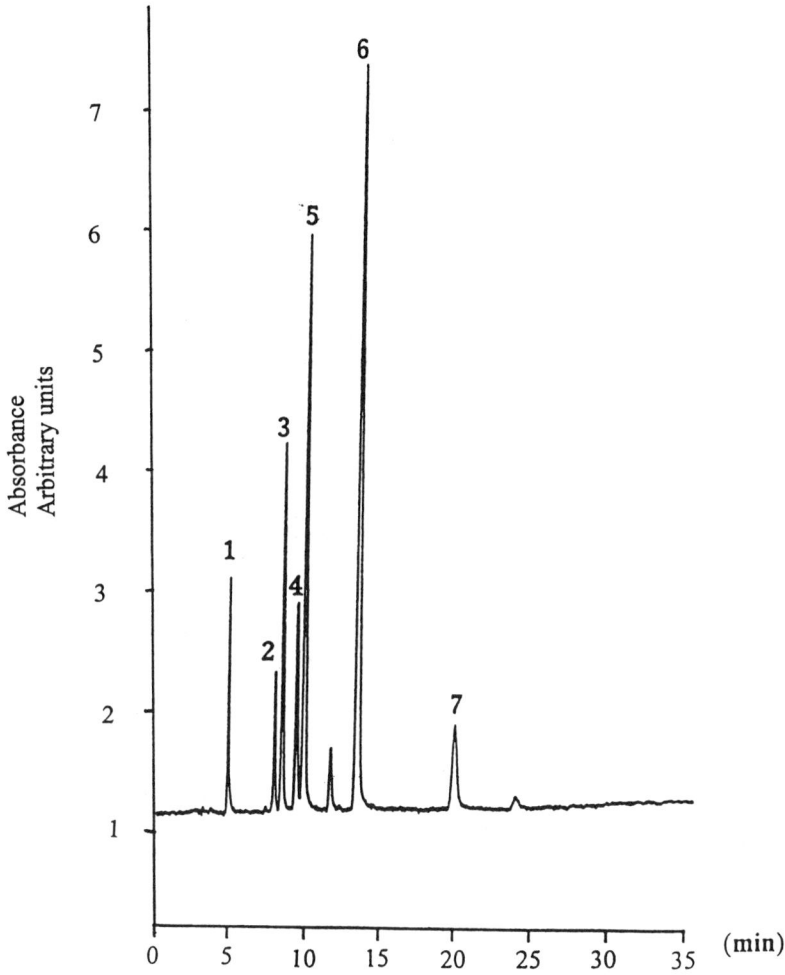

Figure 6. HPLC chromatogram of nitrated polycyclic aromatic hydrocarbons on a packed capillary column (40 cm × 250 μm i.d. packed with 5 μm ODS-1) in the unified chromatograph. Mobile phase acetonitrile/water/methanol (60:30:10 v/v) at room temperature. Detection by UV at 254 nm. Peak identifications: (1) 4-nitroaniline, (2) 1-nitronaphthalene, (3) 2-nitronaphthalene, (4) 2-nitrofluorene, (5) 3-nitrobiphenyl, (6) 9-nitroanthracene, (7) 1-nitropyrene.

(16). A capillary column packed with this material permitted elution of alkanes beyond C_{130} (Figure 7).

To make the best use of packed capillary columns with liquid and supercritical fluid mobile phases, gradient elution is necessary and systems capable of delivering 0.01 to 1.0 µL per minute of modifier are required. We have employed a novel micro SFC-HPLC positive displacement type pump (17) consisting of a pressure control pumping system for SFC and a flow control pumping system for micro HPLC. The SFC pump is designed with dual pump chambers connected in parallel. It provides constant pressure control with continuous flow. The micro LC pump system is designed with dual pumping chambers connected in series. It can be used either as a modifier pump or an isocratic micro LC pump. It provides continuous constant flow control with minimum pressure pulsations.

The pump was very stable under both isocratic and gradient elution conditions, with less than 1% coefficient of variation of retention time in runs in which CO_2 pressure (100 - 300 bar) and methanol modifier content (0.10 - 1.00 µL min^{-1}) were varied over 30 minutes. Figure 8 is the SFC chromatogram of coal tar oil with simultaneous programming of CO_2 pressure and methanol modifier content.

Electrochromatography on packed capillary columns

Recent developments in µHPLC have centred on driving the flow of mobile phase through the columns by means of an electric field rather than by applied pressure (18). This electroosmotic flow (EOF) is generated by applying a large voltage across the column; positive ions of the added electrolyte accumulate in the electrical double layer of particles of column packing, move towards the cathode, and drag the liquid mobile phase with them. As in capillary electrophoresis (CE) and micellar electrokinetic chromatography (MEKC), small diameter (50 - 100 µm) columns with favourable surface area-to-volume ratio are employed to minimise thermal gradients from Ohmic heating, which can have an adverse effect on band widths. Avoiding the use of pressure results in a number of important advantages for CEC over conventional HPLC. Firstly, in HPLC the pressure driven flow velocity, u, through a packed bed depends directly on the square of the particle diameter, (d_p), and inversely on column length, L

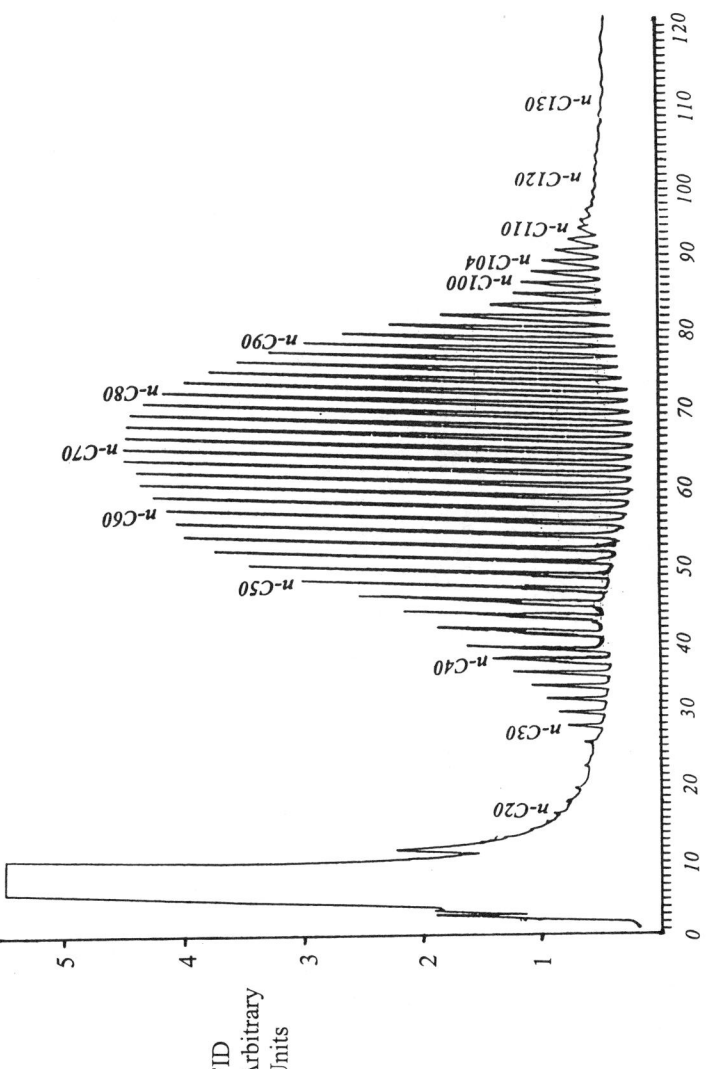

Figure 7. SFC chromatogram of Polywax 1000 polyethylene on a packed capillary column (30 cm × 250 µm i.d. packed with 5 µm hexylsilyl bonded silica) in the unified chromatograph. Mobile phase CO_2 at 120 °C. Pressure programme 200 to 415 bar at 3.5 bar min^{-1}. Detection by FID at 415 °C.

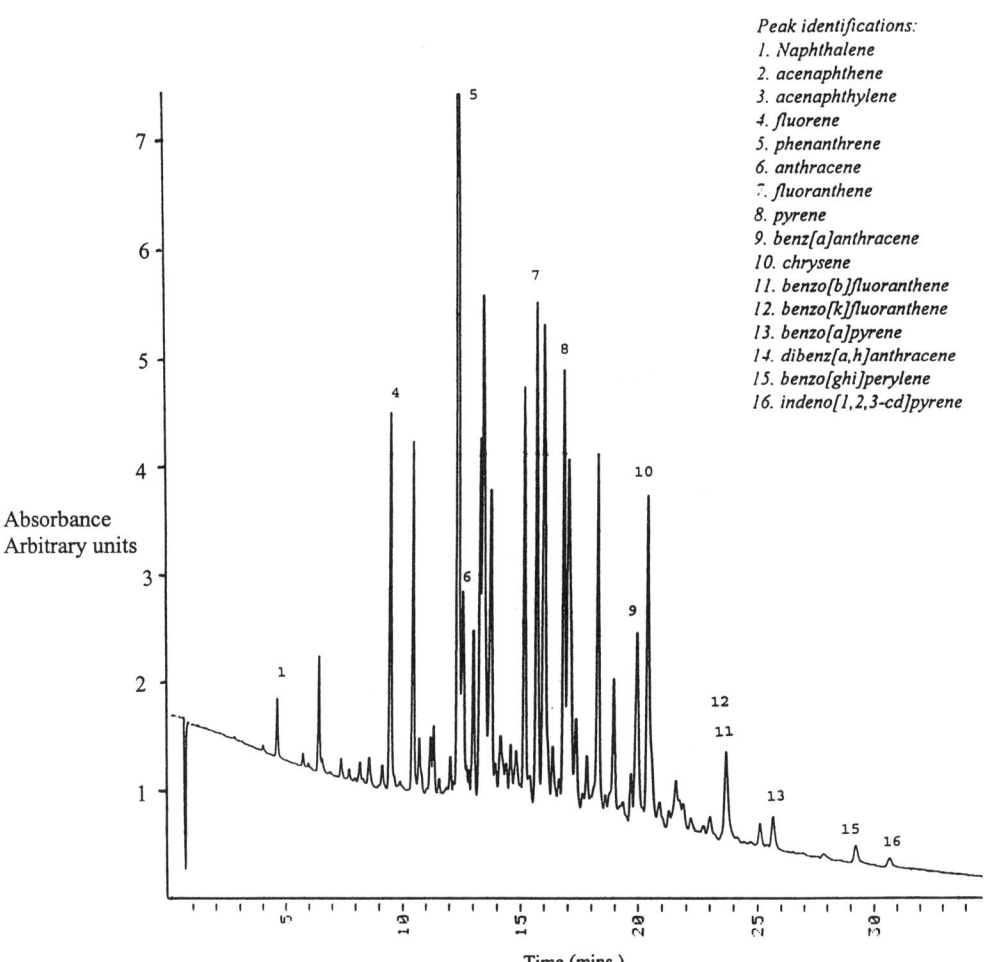

Figure 8. SFC chromatogram of coal tar oil on a packed capillary column (30 cm × 250 μm i.d. packed with 5 μm Spherisorb PAH) in the unified chromatograph. Mobile phase, CO_2 with methanol modifier at 100 °C. Pressure programme 100 to 300 bar CO_2, modifier programme 0.10 to 1.00 μL min^{-1} methanol at 0.03 μL min^{-1}. Detection by UV at 254 nm

$$u = \frac{d_p^2 \Delta p}{\phi \eta L} \quad (1)$$

where Δp is pressure drop across the column, ϕ is column resistance factor and η is mobile phase viscosity. For practical pressures, generally used particle diameters are seldom less than 3 µm, with column lengths restricted to approximately 25 cm. By contrast the electrically driven linear velocity, u_{eo}, is independent of particle diameter and column length so that, in principle, smaller particles and longer columns can be used, provided that the available power supply is sufficient to generate the desired electric field.

Here,

$$u_{eo} = \frac{\varepsilon_o \varepsilon_r \zeta E}{\eta} \quad (2)$$

where ε_o is permittivity of a vacuum, ε_r is mobile phase permittivity, ζ is the zeta potential and E is the electric field.

It follows that considerably higher efficiencies can be generated in CEC than in HPLC. A second consequence of employing electrodrive is that the flow-velocity profile in EOF reduces dispersion of the band of solute as it passes through the column, further increasing column efficiency. The combined effect of reduced particle diameter, increased column length and plug flow leads to high efficiencies, and substantially improved resolution.

Voltages up to 30 kV are supplied to generate the electric field usually for solutions of 1-50 mM buffers in aqueous reverse phase mobile phases; non-aqueous CEC has also been carried out. The dependence of EOF flow rate on solvent dielectric constant has been confirmed, but the electrical potential (the zeta potential) of the boundary between the fixed and diffuse layers (the double layer) of positive ions at the stationary phase wall is less well understood. The conclusion of a theoretical study by Rice and Whitehead which suggested that flat EOF profiles in a capillary of diameter d would result if d were considerably greater than the double layer thickness, δ, has been confirmed by experiment; for channels between particles, however, the influence of δ is less clear. Current indications are that it should be possible to use

Comparison of CEC versus HPLC

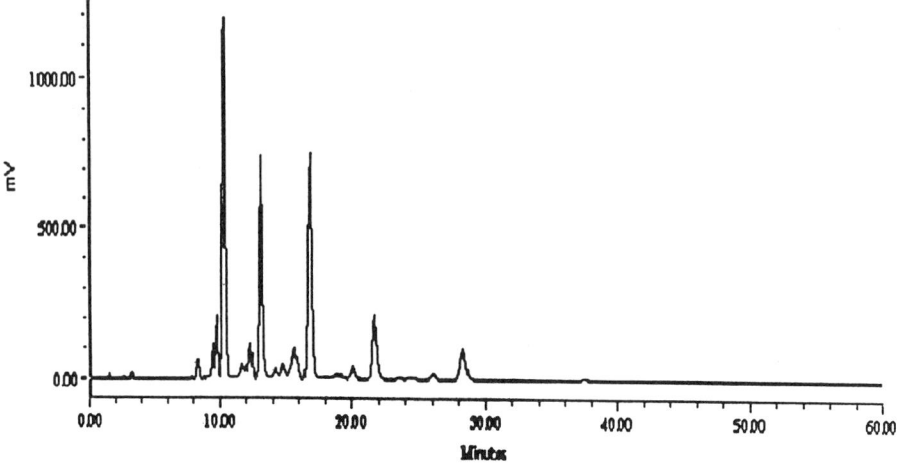

Chromatogram showing HPLC analysis of a C_7-C_9 Phthalate sample.

Mobile phase: 80/20 ACN:TRIS (50mM, pH 7.8), Flowrate: 1 ml/minute,

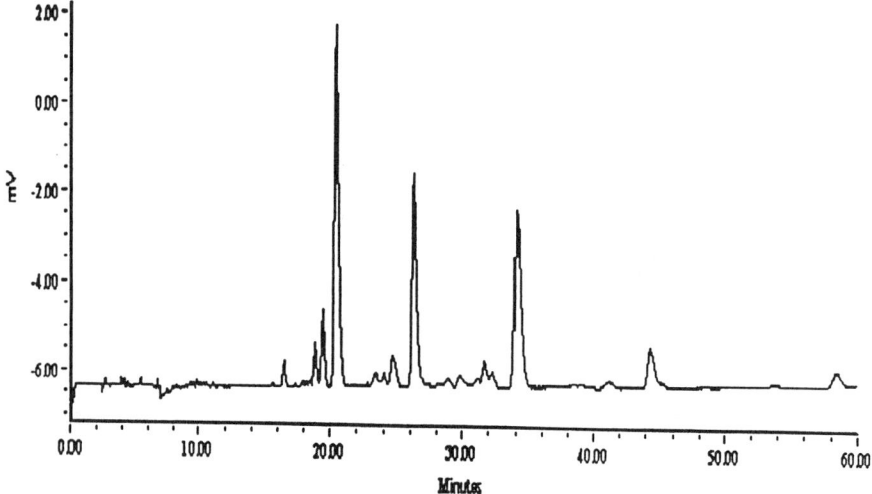

Electropherogram showing CEC analysis of a C_7-C_9 Phthalate sample.

Mobile phase: 80/20 ACN/TRIS (50mM, pH 7.8), Voltage: 30kV,

Figure 9. Chromatograms of a mixture of industrial phthalates (C_7-C_9) obtained by (A) HPLC on conventional 15 cm × 4.6 mm i.d. HPLC column packed with 5 μm Waters Spherisorb ODS-1 and by (B) CEC on a packed capillary column (25 cm × 100 μm i.d.). Mobile phase in both cases 80:20 v/v acetonitrile: TRIS (50 mM, pH 7.8).

monodisperse particles with diameters down to 0.5 μm. Pores sizes of commonly used HPLC particles are too small to give rise to EOF, but larger pore packings show promise. Although CEC has been demonstrated for stationary phases bonded to the walls of open tubes, and in sol-gel derived phases, most work has been carried out on columns packed with HPLC stationary phases (18).

Small diameter columns for CEC are conveniently packed using supercritical CO_2 carrier with ultrasonic agitation of the column. This procedure reproducibly generates columns for reverse phase packings with efficiencies of approximately 250,000 plates per metre for 3 μm particles. A van Deemter plot of h' against u measured on a typical supercritical CO_2 packed capillary is compared with corresponding curves for a C_{18} phase packed capillary operated in HPLC and SFC modes in Figure 3. The h' min value of 1.5 in CEC and the small increase in h' with u are consistent with reduced band broadening from multiple path and mass transfer effects associated with the plug flow of CEC. Such columns generate high resolution reverse phase separations for neutral solutes (Figure 9).

Most CEC work so far has been carried out with stationary phases specifically designed for HPLC work, and a variety of results have been reported from these. Table 2 outlines published results on the separation of PAHs using isocratic CEC. The differences shown, although not normalised in any way, are far greater than one would expect from HPLC comparisons and may result from the packing of these materials into narrow-bore (50-200 μm id) capillaries. Although they may be listed as 3 μm material, all the packings will have unique particle size distributions. In addition, in a manner analogous to the molecular size distribution of polymers, the distribution will vary according to how it is measured; currently there are three ways to characterising particle size distribution, namely number, area and volume. Manufacturers typically do not stipulate which method was used to characterise a particular stationary phase and thus a nominal 3 μm material may vary from company to company. However, extremely noticeable in all of the number distributions is the presence of fine material below 2 μm which is thought to impede the packing process. This material is very difficult to remove *via* the normal air classification used by manufacturers to produce different particle sizes. However, work by some

Table 2: Efficiencies obtained for isocratic CEC of PAHs using HPLC stationary phases

Stationary phase	Range of efficiencies (plates per metre)
3 µm Spherisorb ODS1	200 00 - 240 000
3 µm Nucleosil 100 C_{18}	91 000 - 147 000
3 µm Spherisorb C_{18} PAH	Up to 260 000*
3 µm Synchrom	102 000 - 138 -000
3 µm Vydac C_{18}	160 000
3 µm CEC Hypersil	240 000 - 280 000
*Calculated for a 50 µm id column of length 280 mm, d_p 3 µm and minimum reduced plate height of 1.3.	

manufacturers has led to new particle size distributions that are optimised for the packing procedures used in the packing of 50-200 μm fused silica capillaries.

The formation of bubbles during CEC, which causes serious interference, is due to changes in EOF across the retaining frit which is manufactured *in situ* in the column usually by heating with an electric wire filament. The EOF changes, in turn, result in local changes in liquid velocity and hence in bubble formation. Bubbles can be inhibited by capillary pressurisation and mobile phase degassing, but the most effective procedure is to re-bond octadecylsilyl groups on to the frit to maintain similar EOF velocities to those in the column.

Work has been reported (20) on the use of wide-pore silica materials with pore sizes up to 400 nm as column packings for CEC. These materials are capable of supporting through-particle electroosmosis, and, in principle, of significant increases in efficiency. Wide-pore packings also permit electrically driven size-exclusion chromatography. Thus, Figure 10 shows the separation of a series of narrow molecular-weight polystyrene standards on 30 nm pore size 3 μm silica particles. The retention times of the standards show the well-known logarithmic dependence on molecular weight.

If existing HPLC analyses are to be replaced by CEC methods, the practising analyst must perceive substantial advantages, particularly improved resolution in CEC (Figure 9), along with equivalent performance as regards HPLC variables. Among these are the principle of eluotropy - the correspondence of elution times for different reverse mobile phase compositions. In fact, since u_{eo} depends (equation 2) on the ration ε_r/η for the solvent, very different CEC retention times may be observed for isoeluotropic mobile phases; however, a comparison of retention factors, k, is more appropriate and shows that k values correspond for isoeluotropic solvents in CEC. Further, a well-known feature of HPLC method development, the linear relation between ln k and percentage organic solvent in the mobile phase, is obeyed in CEC for neutral solutes (Figure 11); of course, u_{eo} decreases with increasing percentage organic because of decrease in ε_r/η.

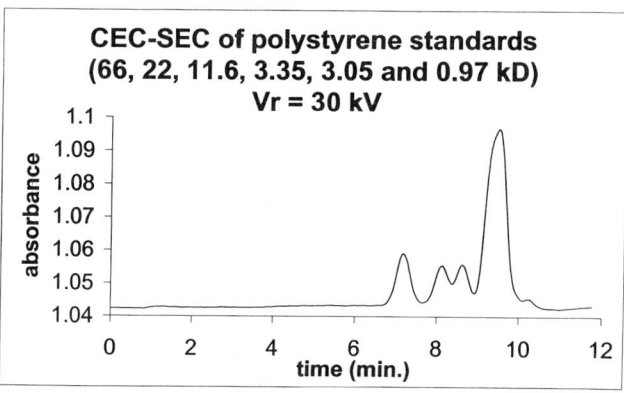

Figure 10. CEC chromatogram of a mixture of polystyrene standards on a packed capillary column (20 cm × 100 μm i.d.) packed with 30 nm pore silica. Mobile phase dimethyl formamide with 0.1 mM tetrabutyl ammonium tetrafluoroborate.

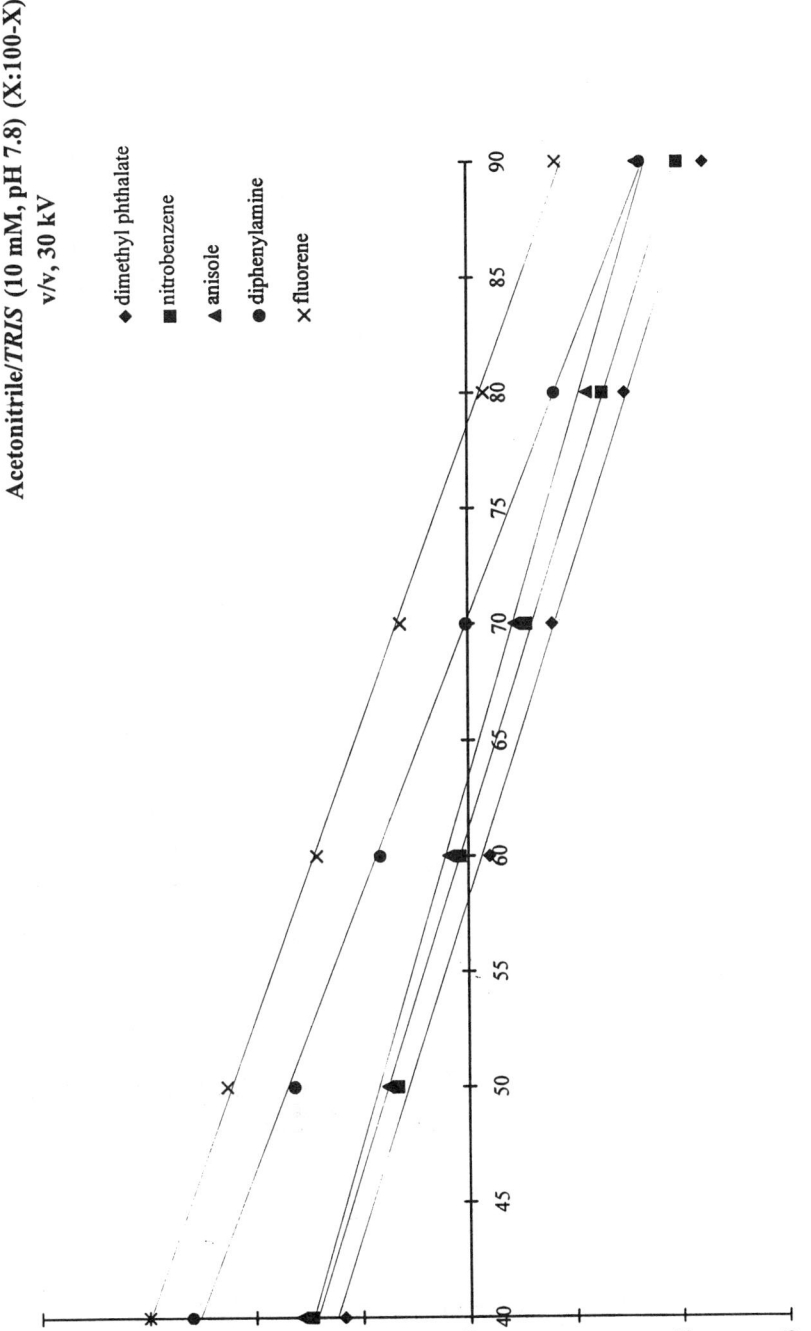

Figure 11. Graph of logarithm of retention factor (k) against acetonitrile content of mobile phase during CEC with acetonitrile/10 mM TRIS, pH 7.8 buffer at 30 kV.

Raman detection - a new strategy for microchromatography

The absence of a sensitive universal detector has greatly limited the types of molecule that can be routinely analysed in HPLC, SFC and GC on packed capillary columns, and an alternative is required to the usual detection methods, UV/visible absorbance and fluorescence, which rely on the presence in the analyte of an appropriate group, or on derivatisation to introduce such a group. Nor do these techniques provide enough structural information to allow molecular identification.

In principle, vibrational spectroscopy should be much more informative in chromatographic detection, providing 'fingerprint' spectra as well as both universal and functional-group selective detection. Infra-red (IR) spectroscopy is, however, severely hampered by complex optics and the necessity of eliminating liquid solvents. For example, we have previously shown (21) how capillary SFC-FTIR is possible with CO_2 mobile phase evaporation, or by use of supercritical xenon as carrier (22), and how µHPLC may be coupled to FTIR by electrospray solvent elimination (23). On the other hand, Raman spectroscopy has considerable advantages over IR in that visible radiation is used, to carry the vibrational information so that silica optics can be used and it may be more easily possible to work in solution and hence avoid solvent elimination. Monochromatic visible photons, usually from a laser, are scattered by vibrating sample molecules and those emerging with lower energy are collected to provide a vibrational spectrum. Normal vibration modes are Raman active if they involve a change of polarizability. Raman spectroscopy should therefore provide both universal and selective detection, and the spectrum will also enable specific identification of separated species.

The historical problem with the Raman technique has been the low efficiency of both scattering and detection, but the detection problem has now largely been solved by technological developments; modern spectrometers can now detect nearly 50% of the Raman scattered photons. Effort can now be concentrated on enhancing the amount of Raman light collected without resorting to molecularly specific resonance processes. Non-universal Raman detectors for HPLC and CE have been demonstrated by other workers (24, 25); these have generally been based on resonance Raman, surface-enhanced resonance Raman, and pre-concentration techniques and have similar limitations to UV absorption and fluorescence detection. Accordingly, we have used a modern Raman spectrometer with high detector efficiency coupled to a microchromatograph for universal and selective detection without pre-concentration or resonance enhancement.

An μHPLC pump was used in isocratic mode to supply mobile phase (methanol/water, or their perdeuterated analogues) to a 250 μm i.d. fused silica column packed with 5 μm Waters Spherisorb ODS-1. The column was joined to a 75 μm or 250 μm i.d. detector cell fabricated from fused silica capillary with the polyimide coating removed. The spectrometer was a Renishaw Raman microscope (26) with a fibre-optic probe attachment (27) (Figure 12). 633 or 515 nm wavelength light from a He Ne or Ar ion laser was passed through a 50 μm diameter optical fibre through line and holographic notch filters, and focused into the detector cell held in a moveable XYZ stage by an X20 microscope objective lens with an approximate depth of field in the capillary of 50 μm and a numerical aperture of 0.46 to maximise collection of Raman scattered light from the solution. The scattered light was collected in the microscope objective and passed through a second fibre to the microscope, dispersed by a diffraction grating, and imaged as the Raman spectrum on to a Peltier-cooled CCD camera. Up to 600 scans per run could be recorded; curves were fitted to all the Raman bands in the spectra, and the area under each curve was calculated to produce a chromatogram.

Deuterated solvents were found particularly useful as the mobile phase, since their Raman bands are shifted to a lower frequency than those of their non-deuterated analogues. For the flow-rates of ~2 μl per minute used here it is quite feasible to use expensive solvents. A comparison of the Raman spectra of nitropropane in CH_3OH/H_2O and in CD_3OD/D_2O shows a much reduced background scattering in the deuterated solvent. Use of deuterated solvents should allow investigation of the C-H stretching vibration for universal detection.

Raman spectra recorded in the range 1250-1700 cm^{-1} at different times during the separation of a mixture of nitrobenzenes are shown in Figure 13. The very strong Raman spectrum of nitrobenzene is observed after 250s followed by the characteristic spectra of 1,3-dinitrobenzene after 360s. A chromatogram (Figure 14) was then generated for an injection of 50 nL of a solution of nitrobenzene and 1,3-

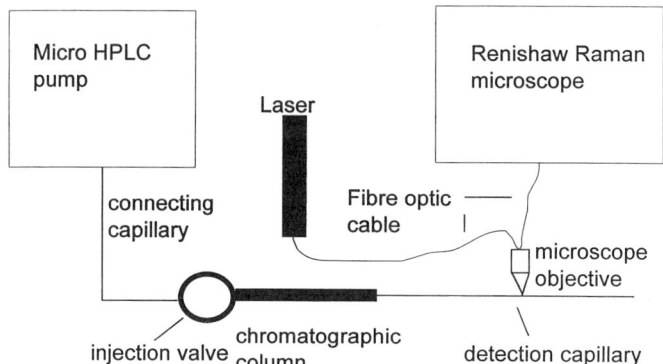

Figure 12. Schematic diagram of micro HPLC chromatograph with microRaman detector.

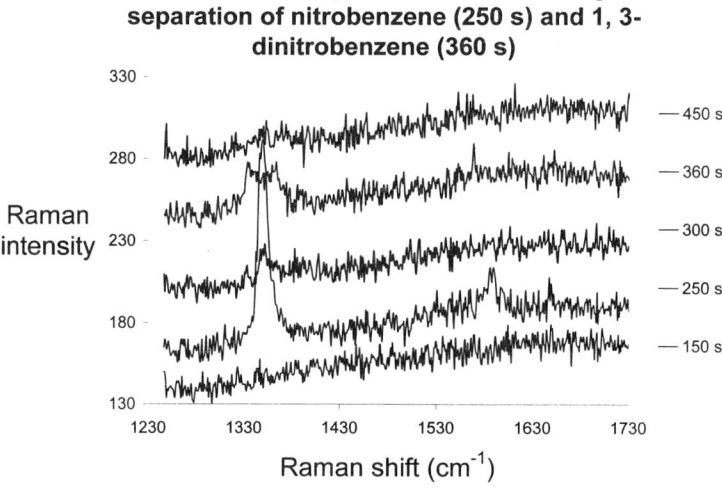

Figure 13. Series of Raman spectra recorded during the HPLC separation of nitrobenzene and dinitrobenzene on a packed capillary column (10 cm × 250 μm i.d. packed with 5 μm ODS-2). Mobile phase CD_3OD/D_2O 75:25.

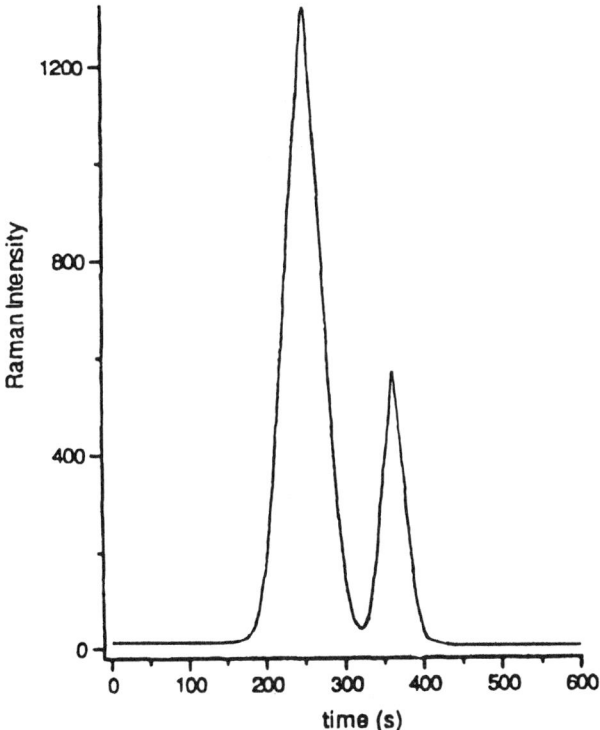

Figure 14. HPLC chromatogram of the summed integrated intensities of the Raman bands of nitrobenzene and 1,3-dinitrobenzene during the chromatographic separation in Figure 13.

dinitrobenzene by integrating the intensities of the bands from each species. The column dead-time could be determined by injection of CH_3OH for which the Raman bands at 1450 and 1462 cm^{-1} could be integrated separately from the 1342 cm^{-1} band of nitrobenzene. A graph of integrated chromatographic peak area against mass of nitrobenzene was linear. A detection limit of 75 ng, corresponding to a signal-to-noise ratio of three, was determined.

Shorter wavelength (λ) laser light improves the sensitivity of the technique since Raman scattering intensity is proportional to λ^{-4}. Thus, we have observed reduced spectral noise and increased response when an Ar ion laser replaced a He-Ne laser. Higher-powered shorter wavelength lasers and improved optics should reduce the current detection limit below 75 ng. In CEC, which is not pressure driven, improved detection limits should be possible by stopped-flow methods in which large numbers of spectra are recorded for a separated mixture component held in the sample cell by switching off the applied voltage.

Work to date suggests that unenhanced Raman detection without preliminary concentration is a promising detector for HPLC on packed capillary columns. Solvent consumption is small and allows the deuterated solvents to be employed. Response is linear, and the current detection limit should be reduced by shorter laser wavelengths, higher power lasers, and improved optics. Current activity is directed towards the wide variety of biological molecules such as applying Raman detection to carbohydrates, proteins and amino acids; these have Raman bands well removed from those of deuterated mobile phases so that derivatisation should not be necessary for detection, and our aim is to develop Raman detection methods for such molecules after HPLC or CE separation with detection limits below 1ng. We are also extending Raman detection to CEC where plug flow, smaller particle diameters and longer columns will reduce band broadening to increase resolution.

Raman detection should also be possible in capillary SFC since the bands from CO_2 are well removed from e.g. the C-H stretch region. Preliminary experiments have allowed us to record spectra of hydrocarbons such as naphthalene in supercritical CO_2 in a 100 μm i.d. capillary.

Acknowledgements

The support of this work by the UK Engineering and Physical Sciences Research Council, by British Petroleum, Renishaw and the Waters Corporation is gratefully acknowledged

References

1. Giddings, J.C. *Dynamics of Chromatography*, Marcel Dekker: New York (1965).
2. Chester, T.L. *Anal.Chem.* **1997**, 69, 165A.
3. Tong, D.; Bartle, K.D.; Clifford, A.C. *J.Chromatogr.A.* **1995**, 703, 17.
4. Tong, D.; Bartle, K.D.; Clifford, A.C. *J.Microcol.Sep.* **1994**, 6, 249
5. Yang, F.J. *Microbore Column Chromatography: A Unified Approach*; Marcel Dekker: New York (1989)
6. Guan, Y.; Zhou, L.; Shang, Z. *J.High Res. Chromatogr.* **1992**, 15, 434.
7. Tong, D; Bartle, K.D; Clifford, A.A.; Edge, A.M.; *J. Microcol.Sep.***1995**, 7, 265.
8. Tong, D; Bartle, K.D; Clifford, A.A; Robinson, R.E. *Analyst*, **1995**, 120, 2461.
9. Myers, M.N; Giddings, J.C. *Anal.Chem.* **1965**, 37, 1453.
10. Robson, M.M.; Dmoch, R.; Bartle, K.D.; Myers, P.; Lynch, T.P. *J.Chromatogr.A.* **1998** submitted.
11. Di Corcia, A. Fritz, D.; Bruner, F. *J.Chromatogr.* **1970** 53, 135.
12. Gidings, J.C.; Schettler, P.D. *Anal.Chem.***1964**, 36, 1483.
13. Huber, J.F.K.; Laver, M.H.; Poppe, H. .*Chromatographia*, **1975**, 112, 377
14. Shen, Y.; Lee, M.L. *Chromatographia*, **1995**, 42, 665.
15. Lee, M.L.; Markides, K.E., *Analytical Supercritical Fluid Chromatography and Extraction;* Chromatography Conferences Inc.: Provo, Utah (1995).
16. Shariff, S.M.; Tong, D.; Bartle, K.D.*J.Chromatogr.Sci.* **1994**, 32, 541.
17. Robson, M.M.; Raynor, M.W.; Bartle, K.D.; Clifford, A.A. *J.Microcol.Sep.* **1995**, 7, 375.
18. Cikalo, M.G.; Bartle, K.D.; Robson, M.M.; Myers, P; Euerby, M.R. *Analyst*, **1998**, 123, 87R.
19. Rice, C.E.; Whitehead, R. *J.Phys.Chem.* **1965**, 69, 4017.
20. Li, D.; Remcho, V.T. *J.Microcol.Sep.* **1997**, 9, 389.
21. Bartle, K.D.; Clifford, A.A.; Raynor, M.W. In *Hyphenated Techniques in Supercritical Fluid Chromatography and Extraction;* Jinno, K. Ed; Journal of Chromatography Library Series No. 53; Elsevier Science: Amsterdam, (1992), p.103.
22. Raynor, M.W.; Shilstone, A.A.; Clifford, A.A.; Bartle, K.D.; Cleary, M.; Cook, B.W. *J.Microcol.Sep.* **1991**, 3, 337.
23. Raynor, M.W.; Bartle, K.D.; Cook, B.W. *J.High Res. Chromatogr.*, **1992**, 15, 361.
24. Pothier, N.J.; Force, R.K. *Appl.Spectrosc.* **1994**, 49, 421.
25. Walker, P.A.; Kowalchyk, W.K.; Morris, M.D. *Anal.Chem.* **1995**, 67, 4255.
26. Williams, K.J.P.; Pitt, G.D.; Smith, B.J.E.; Whitley, A.; Batchelder, D.N. *J.Raman Spectrosc.* **1994**, 25, 131.
27. Howard, I.P.; Kirkbride, T.E.; Batchelder, D.N.; Lacey, R.J. *J.Forensic Sci.* **1995**, 40, 883.

Chapter 10

Applications of Enhanced-Fluidity Liquid Mixtures in Separation Science: An Update

S. V. Olesik

Department of Chemistry, Ohio State University, 100 West 18th Avenue, Columbus, OH 43210

Enhanced fluidity liquid mixtures can provide the positive attributes of supercritical fluids and polar liquids to separation science. The physicochemical properties of enhanced-fluidity liquid solvents are described. A range of possible applications to separation science is highlighted. These include reversed-phase, normal-phase liquid chromatography, size exclusion chromatography, critical chromatography and extraction chemistry.

Enhanced-fluidity liquids are mixtures of common liquid solvents such as alcohols, water, tetrahydrofuran, and hexane combined with a liquified gas such as carbon dioxide. These liquid mixtures share the positive attributes of both supercritical fluids and liquids. The fast mass transport characteristics and the control of solvent properties with pressure and temperature variation are attributes of supercritical fluids that can improve a separation process. However, the commonly-used supercritical fluids are only moderately polar even when cosolvents are added. Many of the compounds encountered in separation science are highly polar. Therefore, the use of supercritical fluids in separation science is limited in its scope. Alternatively, there are many liquids such as alcohols, water and others that provide adequate polarity to dissolve and separate highly polar solutes. However, the diffusion coefficients in liquids are commonly two orders of magnitude lower than those of supercritical solvents and their viscosities are also significantly higher. Enhanced-fluidity liquid mixtures represent a portion of the phase diagram that may be used to combine the useful attributes of supercritical fluids and liquids. This chapter will review some of the physicochemical properties of enhanced-fluidity liquid mixtures, describe some of the applications of enhanced-fluidity liquids in separation science and provide an update on more recent efforts toward expanding the uses of enhanced-fluidity liquids.

Physicochemical Properties of Enhanced-Fluidity Liquid Mixtures

Solvent-Strength. Kamlet-Taft solvatochromic parameters (π^*, α, and β) have been used to evaluate the change in polarity of mixtures as a function of added liquefied gas. π^* measures the dipolarity and polarizability of the solvent, α measures the hydrogen-bond donating capability, and β measures the hydrogen-bond accepting capability (*1*). These parameters are measured using the shift in the UV/visible absorbance spectrum of selected solvatochromic indicators. Solvatochromic indicators are useful for the measurement of the solvent polarity because the shift in the absorbance bands depends on the composition of the solvation sphere that envelopes the indicator. Kamlet-Taft parameters have advantages over other solvatochromic parameters in that they are able to separate out different types of intermolecular interactions.

The variation of the solvent strength of methanol/CO_2 mixtures (25 °C and 170 atm) is shown in Figures 1 and 2.(*2*) The Kamlet -Taft α and β parameters of the methanol/CO_2 mixtures are approximately the same as that of methanol until more than 0.60 mol fraction CO_2 is added (Figure1). The π* parameter decreased more so with added CO_2. The addition of 0.60 mole fraction CO_2 caused a 50% decrease in the π* value (Figure 2). Park *et al.* (*3*) measured the variation in the Kamlet-Taft α parameter for mixtures of H_2O and organic solvents. Interestingly the shape of the curve in Figure 1 is similar to their observations when H_2O was mixed with a solvent with no H-bond acidity, such as tetrahydrofuran. For the H_2O/THF mixtures, the α parameter decreased slightly with addition of more organic solvent, but the mixture maintained substantial acidity up to volume fractions of THF as high as 0.90 (*3*).

The variation of the solvent strength of methanol/H_2O/CO_2 mixtures as a function of added CO_2 was also measured (*4*). The Kamlet - Taft π*, and α values decreased by approximately 15% when 30% CO_2 was added to a 0.70/0.30 mole ratio methanol/H_2O mixtures. However, interestingly, the β parameter increased continuously with the addition of CO_2 by approximately 25%.

Similar measurements of Kamlet-Taft β, and π* parameters of THF/CO_2 mixtures were also measured for the THF/CO_2 mixtures at 26 °C and 136 atm. The α parameter was not measured because THF has no measurable H-bond acidity. The value of the π* parameter decreased with added CO_2. For mole fractions of CO_2 greater than 0.70, π* decreased drastically with added CO_2. Similarly the measured β parameter decreased with added CO_2 with a more drastic decrease occurring for compositions greater than 0.50 mole fraction CO_2.

Diffusion Coefficients. An understanding of diffusion rates in high pressure liquid mixtures is required when these mixtures are used as mobile phases for chromatographic separation or as solvents for extraction. Diffusion rates are most commonly described by comparing the diffusion coefficients of solutes at infinite dilution. The diffusion coefficients of solutes in enhanced-fluidity liquid mixtures have been determined for a number of solutes in various enhanced-fluidity liquid mixtures using the chromatographic band dispersion technique.

The variation of diffusion coefficients for benzene in methanol/CO_2 mixtures at 25 °C and 170 atm across the entire composition range (0-100 mol% CO_2) was measured (*5*). By the addition of 60 mole% CO_2 to methanol the diffusion coefficient of benzene increased by approximately100%.

Figure 3 compares the diffusion coefficients of styrene in pure THF, 80/20 mol% and 60/40 mol% THF/CO_2 over the temperature range of 24 to 80 °C at 170 atm (*6*). The

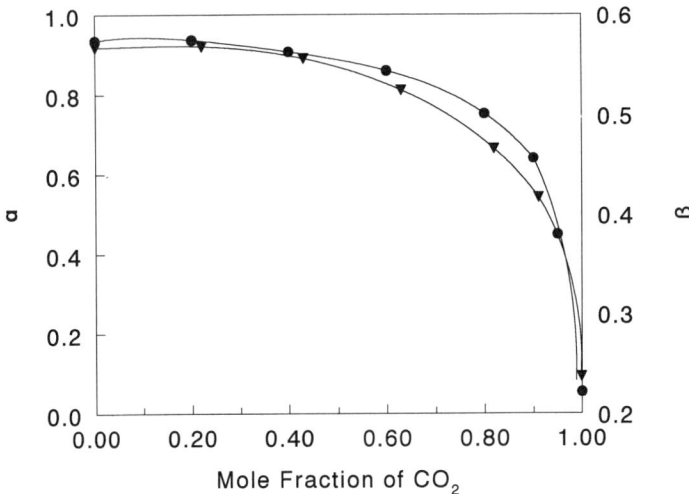

Figure 1. Variation of α (●) and β (▼) as a function of added CO_2 to methanol at 25 °C and 170 atm. (Reproduced with permission from ref 5. Copyright 1990, American Chemical Society)

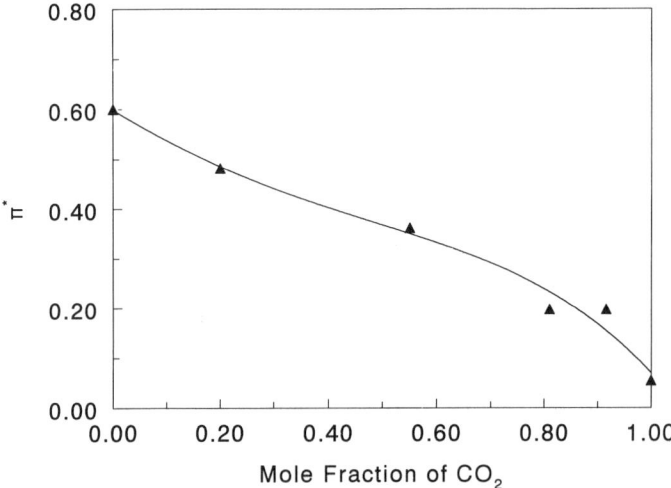

Figure 2. Variation of π^* as a function of added CO_2 to methanol at 25 °C and 170 atm. (Reproduced with permission from ref 5. Copyright 1990, American Chemical Society)

diffusion coefficient increased by approximately 70 % over this temperature range for the solvent mixture studied. The combination of added CO_2 and a temperature increase caused the greatest increase in the diffusion coefficient. The diffusion coefficient of styrene in 60/40 mol% THF/CO_2 at 80°C was approximately 150% greater than that in pure THF at 24°C.

Figure 4 compares the diffusion coefficients of anthracene in ternary mixtures of ethanol/H_2O/CO_2 (7). For a 0.61/0.39 mole ratio ethanol/H_2O mixture at 25°C, the diffusion coefficient increased by 200% when 40 mol% CO_2 was added. By increasing the temperature of the 0.61/0.39 mole ratio ethanol/H_2O mixture to 60°C, the diffusion coefficient increased by approximately 130% and the combination of adding 40 mol% CO_2 and increasing the temperature to 60°C caused an increase of 460%.

Similar studies were also undertaken using methanol/H_2O/CO_2 mixtures over the temperature range of 25-120°C (8). For a 0.70/0.30 mole ratio methanol/H_2O mixture at 24°C the addition of 30 mol% CO_2 caused a 100% increase in the diffusion coefficient. In comparison, increasing the temperature to 60°C caused approximately the same change in the diffusion coefficient of benzene as the addition of 30 mole % CO_2 to the 70/0.30 mole ratio methanol/H_2O mixture. By combining a temperature increase of 60°C and the addition of 30 mole% CO_2 the diffusion coefficient increased by 750%. Interestingly, the addition of 30 mole % CO_2 combined with the increase in temperature to 60°C provides diffusion coefficients of benzene that are comparable to those in CO_2 at 24°C and 160 atm (9). Clearly, when high solute diffusivity is desired, the combination of adding a liquefied gas and increasing the temperature may be the optimum choice.

Equation 1 shows an empirical expression that illustrates the relationship between the diffusion coefficient of a solute and the viscosity of the solvent:

$$D \eta^p = AT \qquad (1)$$

where D is the diffusion coefficient, η is the viscosity of the solvent, T is absolute temperature and A and p are constants that depend on the radii of the solute molecule (10). According to equation 1, if the solvent properties cause the diffusion coefficient of the solute to increase, the viscosity of the solvent decreases. The viscosity of methanol/CO_2 mixtures at 25°C and 170 atm was measured. The viscosity decreased by approximately an order of magnitude across the composition range of 0-100% CO_2. With 60 % CO_2 in the mixture the viscosity decreased by 70%. Therefore, the addition of liquefied CO_2 also substantially decreases the viscosity of the solvent.

In summary, enhanced-fluidity liquid mixtures have mass transfer properties that approach those of supercritical solvents. In addition, the high solvent strength of the organic solvents used in the mixtures is predominately retained even when as much as 50-60 % liquefied gas is added to the mixture. These attributes can be used to advantage in various areas of separation science.

Enhanced-Fluidity Liquid Chromatography

Band dispersion in liquid chromatography is typically controlled by the diffusion into and out of the stagnant mobile phase that is found in the pores of the packing material and by

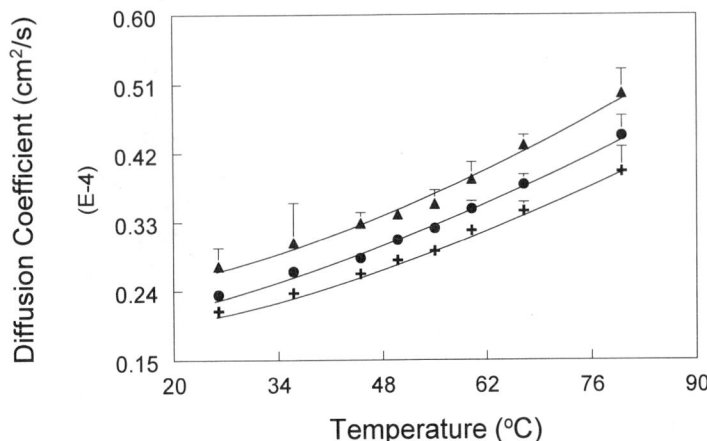

Figure 3. Variation of diffusion coefficients of styrene with temperature for different THF/CO_2 mixtures: (+) THF, (●) 80/20 mol% THF/CO_2, (▲) 60/40 mol% THF/CO_2. (Reproduced with permission from ref. 12. Copyright 1997, Elsevier)

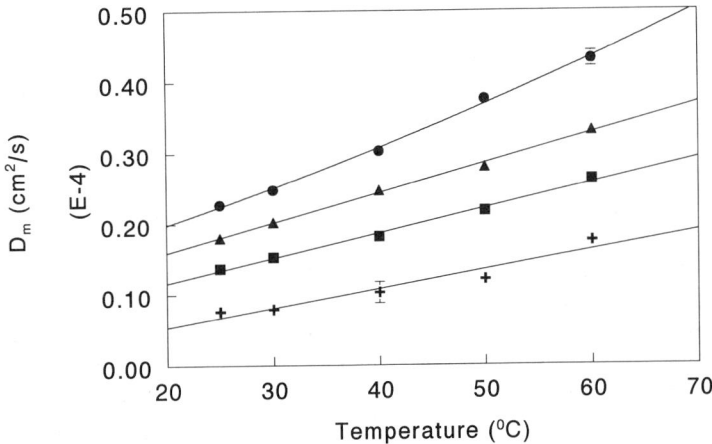

Figure 4. Variation of diffusion coefficients of anthracene as a function of added CO_2 to a ethanol/H_2O (0.61/0.39) mixtures and as a function of temperature. (+) 0, (■) 20, (▲) 30, (●) 40 mol% CO_2. (Reproduced with permission from ref. 7. Copyright 1998, American Chemical Society).

the radial diffusion across the laminar flow profile in the mobile phase. Both contributions to band dispersion are inversely related to the diffusion coefficient of solvent and linearly related to the retention factor. Therefore increased diffusion coefficients in enhanced-fluidity solvents should result in decreased band dispersion. Accordingly, increased efficiency has been observed for normal-phase and size exclusion chromatography (6), (11), (12)). Therefore the decrease in solvent strength as a function of added CO_2 will also cause the retention factor to decrease which will also result in lower band dispersion. For example at a reduced velocity, v, of 10, the addition of 0.5 mole % CO_2 to a methanol/H_2O mixture using a C-18 stationary phase causes a change in reduced plate height from 11 to 4. Substantial improvements in efficiency for reversed-phase separations using enhanced-fluidity mixtures have been observed.

The optimum linear velocity also shifts to higher values when the diffusion coefficients increase. Therefore the speed of analysis increases substantially for separations using enhanced-fluidity solvents as mobile phase. For example, in reversed-phase separations the combination of increased diffusion coefficients and decreased retention factors decreased the separation time for a sample of eight polynuclear aromatic hydrocarbon from 42 minutes to15 minutes by adding 30% CO_2 to the 0.70/0.30 mole ratio methanol/H_2O mixture (8).

Due to the small pressure drop across packed columns when enhanced-fluidity liquid mixtures are used as eluents, highly efficient separations are possible by placing columns in series. For example, the efficiency of the chromatographic systems was quadrupled for the separation of a coal tar sample by placing in series four C-18 columns in series (13).

When CO_2 is added to a mobile phase that contains H_2O, carbonic acid is formed. In reversed-phase chromatography, pH control is important when solutes are ionizable. The retention of the ionizable solutes is highest when the pH is controlled to maintain their neutrality. Two possible solutions to this problem are under study. The problem is completely eliminated if another liquefied gas is used that does not dissociate in water. Fluoroform is a reasonable alternative. It has a low critical temperature and pressure (26.2°C and 25.9 atm) and its viscosity is lower than that of CO_2 under similar conditions(14). An initial study using fluoroform involved the separation of triazine herbicides and their metabolites (15). Figure 5 shows a comparison of band dispersion as a function of linear velocity for atrazine using 64/36 mol% methanol/H_2O, 51/29/20 mol% methanol/H_2O/CO_2 and 51/29/20 mol% methanol/H_2O/CHF_3 as the mobile phase. The addition of fluoroform provided the least band dispersion. The other solution to the problem of mobile phase acidity control is to use buffers. By forming buffers in the enhanced-fluidity mobile phase the pH can more readily be controlled. Phosphate buffers with pH = 7.0 were added to the same methanol/H_2O, methanol/H_2O/CO_2, methanol/H_2O/CHF_3 mobile phases. Figure 6 compares the band dispersion for atrazine as a function of linear velocity with buffer. The addition of the phosphate buffer improved the band dispersion using all three mobile phases. However, the mobile phase containing fluoroform caused the least band dispersion. To utilize buffers effectively in enhanced-fluidity liquids, the apparent pH of the mobile phase should be known. These measurements are underway for buffers with a range of pK_as.

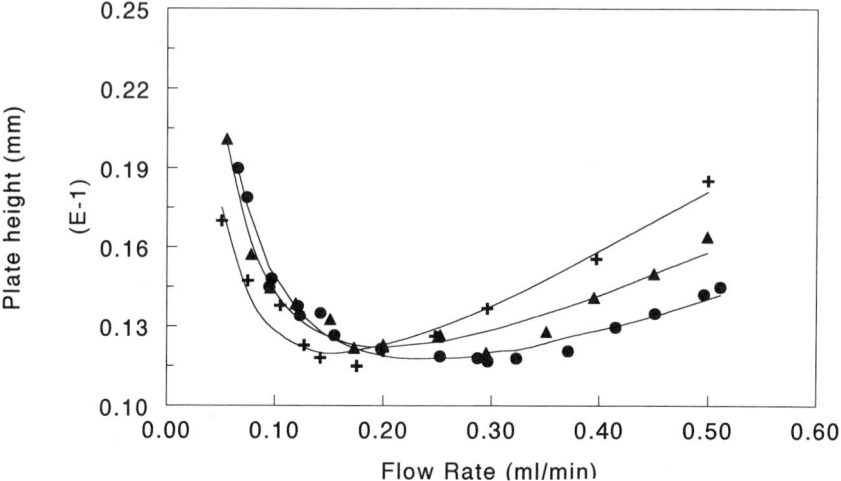

Figure 5. Variation of reduced plate height for atrazine with mobile phase flow rate at 238 atm for different mobile phase compositions (+) 64/36 mol% methanol/H_2O; (▲) 51/29/20 mol% methanol/H_2O/CO_2; (●) 51/29/20 mol% methanol/H_2O/CHF_3. (Reproduced with permission from ref 15. Copyright 1998, American Chemical Society).

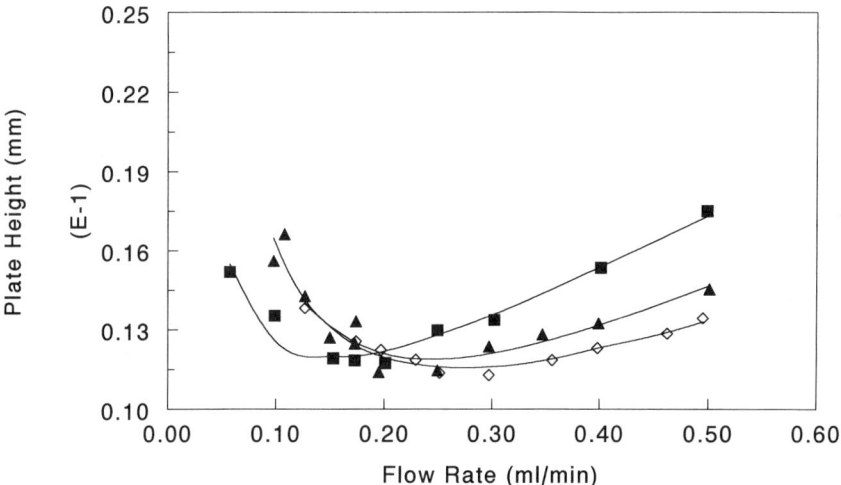

Figure 6. Variation of reduced plate height for atrazine with mobile phase flow rate at 238 atm for different mobile phase compositions (■) 64/36 mol% methanol/10 mM phosphate buffer; (▲) 51/29/20 mol% methanol/10 mM phosphate buffer/CO_2; (◇) 51/29/20 mol% methanol/10 mM buffer/CHF_3. (Reproduced with permission from ref. 15. Copyright 1998, American Chemical Society).

Critical chromatography is another separation method that can be improved by using enhanced-fluidity liquid solvents. This is a type of chromatography that was first described by Gorshkov et al.(16)(17). It is a powerful tool for determining the functionality distribution in telechelic polymers (functionalized polymers) and the distribution of polymers in copolymers. If the functionality distribution of a polymer is to be determined, then first the critical condition of the polymer backbone is found. Next, at the critical condition for the polymer backbone, the functionalized polymers are separated. This separation will be based completely on the differences in Gibbs free energy of transfer between the stationary phase and the mobile phase for the different functionalities. At the critical condition, the change in free energy associated with the transfer of the polymer into the stationary phase is zero (enthalpic interaction is exactly compensated by entropic interaction). Therefore the retention volume of the polymer does not depend on molecular weight at the critical condition (i.e all nonfunctionalized polymer elutes at the same retention volume).

The most common method of finding the critical condition for a specific polymer is to mix a good solvent and a nonsolvent for the polymer and vary the relative proportions of each until the critical condition is found. With common liquid solvents, this is sometimes a difficult task and for some polymers it is difficult to find the critical condition. To maintain the critical condition the volume ratio of the liquids in the mixtures must be maintained, in some cases, to accuracies as small as 0.1 % v/v. The accuracy of most HPLC mixing systems for pumps is typically 1%. Therefore it is often difficult to maintain the appropriate mixture with the HPLC instrumentation unless solvents are premixed.

Temperature variation is the only other variable that can be used with common organic solvents to approach the critical condition. For low molecular weight solutes an increase in temperature decreases the retention since adsorption is an exothermic process. However, for the higher molecular weight polymers using temperature variation to find the critical point is more problematic. An increase in temperature can actually cause an increase in retention.

We have initiated studies on the use of enhanced-fluidity liquid mixtures as solvents for critical chromatography. We can change the solvent strength of these mixtures with high precision by simply varying the pressure and temperature of the solvent. Our use of enhanced - fluidity liquids has added substantial flexibility to critical chromatography. Figure 7 illustrates that for a given polymer system, a size-exclusion separation, a critical condition separation and liquid adsorption chromatogram can be obtained using the same mobile phase by changing the pressure of the mobile phase (18). In this same study the critical condition for polystyrene standards was compared using mixtures of THF/acetonitrile, THF/methanol and, THF/CO_2. The largest molecular weight polystyrene polymer that had the same approximate critical condition as the lower molecular weight solutes was much larger (M.W. ≈100,000) using THF/CO_2 mixtures than THF/acetonitrile or THF/methanol (M.W.≈20,000). The use of CO_2 as a cosolvent clearly expanded the useful molecular weight range for the critical condition for polystyrene. Similar results have been observed for other polymers as well.

When a telechelic polymer is separated at the critical condition of the polymer backbone, the separation is based on the weak interactions of sometimes a single functionality with the stationary phase. Increased selectivity and/or increased efficiency are therefore highly desirable when working at the critical condition. Our initial efforts involved improvement of the efficiency of the separation. Because enhanced-fluidity liquids have low viscosities, markedly longer columns can be produced and used than is possible with conventional liquids. For example, two-meter long packed-capillary columns were packed and used for critical chromatography. These columns had efficiencies as high as 60,000 - 100,000 plates. These long packed-capillaries provided separations of functionalized polymers or copolymers that were not possible with analytical-scale columns even when they were coupled in series. Figure 8 shows the separation of carboxy and dicarboxy-terminated polystyrene standards with the same molecular weight (MW_w = 50,000) using a 1.8- meter x 250-μm i.d. column that was packed with 5μm silica particles(*19*). More work is in progress to expand the range of applications of critical chromatography.

Enhanced-Fluidity Liquid Extraction

The fast diffusivity, low viscosity and high solvent strength of enhanced-fluidity liquid mixtures can be very useful for the extraction of polar solutes from highly adsorptive surfaces. The use of CO_2, methanol/CO_2, and methanol/H_2O/CO_2 under supercritical and enhanced-fluidity liquid conditions for the extraction of a range of polar pollutants from sediment and dust samples was compared (*20*) (*21 - 26*). Quite often the extraction solvent that provided the highest extraction yields was a liquid mixture containing significant proportions of H_2O.

For example, the extraction of eleven different substituted phenols and nitrophenols from house dust was compared using CO_2, 20/80, 30/70 and 40/60 mol% methanol/CO_2 and 9.0/1.0/90, 16.9/3.1/80.0, 25.4/ 4.6/70, 32.1/7.9/60.0 and 40.1/9.9/50.0% mole % methanol/H_2O/CO_2. Statistical analysis of the results showed that the highest extraction yields were obtained using methanol/H_2O/CO_2 mixtures containing 3.1 - 7.9 mol % H_2O. Also, the extraction kinetics using enhanced-fluidity liquid mixtures were similar to those found when using supercritical conditions.

Summary and Future Developments

As mentioned in other chapters of this book, useful chromatographic separations can be obtained by using various regions of the phase diagram for the mobile phase. The choice of mobile phase composition, pressure and temperature should be based on the desired performance of the chromatographic system.

We have studied the use of high fluidity liquids for a range of chromatographic techniques. Significant gains in performance were observed in all of the studied techniques (normal-phase, reversed-phase, size exclusion and critical chromatography). Our studies have not been exhaustive. However to date, the most significant gains were found in reversed-phase chromatography and critical chromatography. Accordingly, we continue to pursue those areas of separation science. Because reversed-phase chromatography is most commonly used with significant proportions of H_2O in the mobile phase, we continue to seek enhanced-fluidity liquid mixtures that can accommodate large proportions of H_2O.

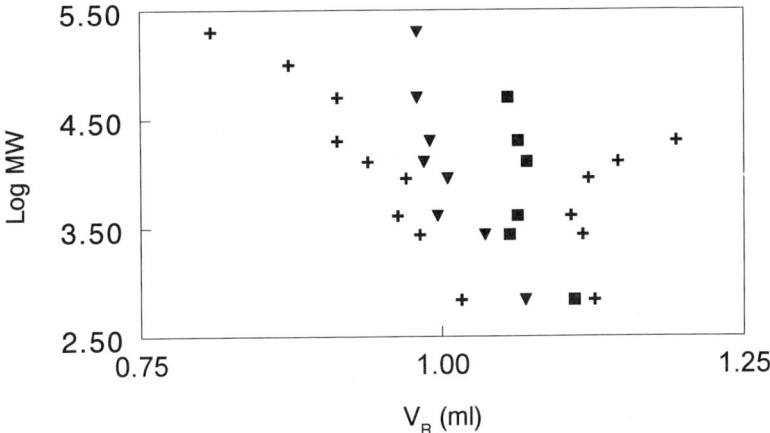

Figure 7. Effect of pressure variation on calibration curve for polystyrene standards using 53.7 mol% CO_2 in THF (+) 69.3, (■) 92.3, (▼) 116, (●) 136 atm. (Reproduced with permission from ref. 18. Copyright 1998, American Chemical Society).

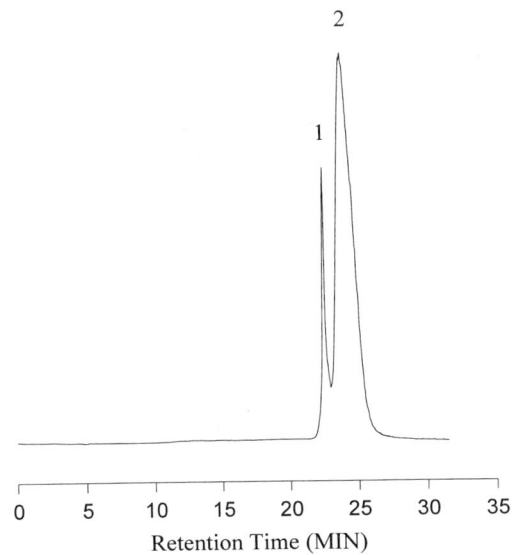

Figure 8. Chromatogram of di (1) and mono carboxy (2) terminated polystyrene (MW = 50,000) at the critical condition for polystyrene. Conditions were: 1.8 m × 250 μm i.d. silica (200 Å pore size, 5 μm particle size) packed column; 54% CO_2 in THF mobile phase; column pressure 3700 psi at 70 °C using an evaporative light scattering detector. (Reproduced with permission from ref. 19. Copyright 1998, American Chemical Society).

chromatography). Our studies have not been exhaustive. However to date, the most significant gains were found in reversed-phase chromatography and critical chromatography. Accordingly, we continue to pursue those areas of separation science. Because reversed-phase chromatography is most commonly used with significant proportions of H_2O in the mobile phase, we continue to seek enhanced-fluidity liquid mixtures that can accommodate large proportions of H_2O. The best combination that we have found to date for reversed-phase work is methanol/fluoroform mixtures. Almost 50 mole % H_2O can be included in these mixtures (27). Highly polar solutes dissolve in these mixtures. We are beginning to evaluate buffering conditions for enhanced-fluidity liquids. To date, we have generated solutions using methanol/H_2O/CO_2 mixtures with apparent pH values ranging from approximately 2 - 7 (28). Recently, we have found that the use of buffered, enhanced-fluidity liquid mixtures as mobile phases often improves significantly the separation of chiral compounds (29).

LITERATURE CITED
1. Taft, R.W.; Kamlet, M. J. *J. Am. Chem. Soc.* **1976**, *98*, 2886.
2. Cui, Y.; Olesik, S.V. *Anal. Chem.* **1991**, *63*, 1812.
3. Park, J.H.; Jang, M.D.; Kim D.S.; Carr, P. W. *J. Chromatogr. A* **1990**, *513*, 107.
4. Cui, Y.; Olesik, S. V. *J. Chromatogr. A* **1995**, *691*, 151.
5. Cui, Y.; Olesik, S. V. *Anal. Chem.* **1991**, *63*, 1812.
6. Yuan, H.; Olesik, S. V. *J. Chromatogr. A* **1997**, *785*, 35.
7. Souvignet I.; Olesik, S. V. Anal. Chem. **1998** *70*, 2783.
8. Lee, S.T.; Olesik, S. V. *Anal. Chem.* **1994**, *66*, 4498.
9. Sassiat, P. R.; Mourier, P.; Caude, M. H.; Rosset, R. H. *Anal. Chem.* **1987**, *59*, 1164.
10. Chen, S.-H.; Davis, H. T.; Evans, D. F. *J. Chem. Phys.* **1982**, *77*, 2540.
11. Lee, S. T.; Olesik, S. V. *J. Chromatogr. A* **1995**, *707*, 217.
12. Yuan, H.; Olesik, S.V. *J. Chromatogr. A* **1997**, *785*, 35.
13. Lee, S. T.; Olesik, S. V.; Field, S. M. *J. Microcol. Sep.* **1995**, *7*, 477.
14. Handbook of Chemistry at Physics, 72nd Ed.; CRC Press; Boca Raton, FL 1991.
15. Yuan, H.; Olesik, S. V. *Anal. Chem.* **1998**, *70*, 1595.
16. Gorshkov, A. V.; Verenich, S. S., Evreinov, V. V., Entelis, S. G. *Chromatographia* **1988**, *26*, 338.
17. Entelis, S. G.; Evreinov, V. V.; Gorshkov, A. V. *Adv. Polym. Sci.*, **1987**, *76*, 129.
18. Souvignet, I.; Olesik, S. V. *Anal. Chem.* **1997**, *69*, 66.
19. Yun, H.; Olesik, S. V. *Anal. Chem.* **1998**, *70*, 3298.
20. Kenny, D; Olesik, S. V. *J. Chromatogr. Sci.* **1998**, *36*, 66.
21. Kenny, D; Olesik, S. V. *J. Chromatogr. Sci.* **1998**, *36*, 59.
22. Yuan, H.; Olesik, S. V. J. *J. Chromatogr. A* **1997**, *764*, 265.
23. Monserrate, M.; Olesik, S. V. *J. Chromatogr. Sci.* **1997**, *35*, 82.

Chapter 11

Universal Chromatography for Fast Separations

Milton L. Lee and Christopher R. Bowerbank

Department of Chemistry and Biochemistry, Brigham Young University, P.O. Box 25700, Provo, UT 84602–5700

Fast separations are becoming increasingly important in all forms of chromatography, and considerable research is underway to develop instrumentation and column technology that can perform fast separations in a reliable manner. In truly high-speed chromatography, temperature and composition gradient programming are not practical, primarily because of the relatively long times required for re-equilibration. These and other requirements of fast separations lead to the strongest case for universal chromatography, where "universal" implies that the mobile phase can be in any form (i.e., gaseous, supercritical fluid, or liquid) at any point in the chromatographic column, and the instrumentation, including the column, can be used for all forms of column chromatography. In this paper, coupled, unified, and universal approaches to chromatography are reviewed, and universal instrumentation and column technology are described, particularly applicable to fast chromatography. This paper is not intended to be a comprehensive review of the literature, but a general overview illustrated with selected examples.

Traditionally, column chromatography has been divided into the main classifications of gas chromatography (GC) and liquid chromatography (LC). In recent years, supercritical fluid chromatography (SFC) has taken its place between the two main chromatographic techniques because the properties of supercritical fluids lie between those of gases and liquids. SFC is truly intermediate between GC and LC, and it tends to more closely resemble one or the other in theory, instrumentation, and performance when operated at high pressure (i.e., LC) or low pressure (i.e., GC). The traditional boundaries between GC and LC have become even more blurred in recent years with the introduction of new techniques such as elevated temperature liquid chromatography (ETLC), enhanced fluidity liquid chromatography (EFLC), and solvating gas chromatography (SGC). These new forms of chromatography have all aimed for the advantages claimed by SFC, better solute diffusion and lower viscosity than experienced with typical liquid mobile phases, while at the same time retaining mobile phase solvating power.

One practical characteristic common to these intermediate chromatographic techniques is that pressure control of the mobile phase in the column is required. This was predicted by Myers and Giddings (1) in 1966 to be useful, when they commented that, "Gas and liquid chromatography are tools having many experimental dimensions. One of the least-explored dimensions — a dimension for which we can conceive enormous range and influence — is pressure." In fact, pressure programming is used as the primary method of controlling elution in SFC and SGC. It is interesting to note here that even GC and LC require high pressures when high speed analysis is desired. In fact, additional universal requirements, i.e., smaller diameter columns, smaller diameter packing materials, and lower dead-volume connectors, to name a few, for all of the chromatographic methods when high speed is desired tend to bring even more universality to the instrumentation and practice of the various techniques.

In the following sections of this chapter, different approaches to combining the various forms of chromatography together, with the desire to unify these forms in some way, are briefly reviewed. The concept of universal chromatography, which includes the fundamentals of unified chromatography, is also described along with representative chromatographic examples. High speed separations present the best conditions for universal chromatography, and the rationale is discussed. The classical van Deemter equation for chromatographic efficiency was converted into a more general (universal) form to account for compressible mobile phases.

Universal Instrumentation

One way in which the chromatographic methods can be unified is by developing common instrumentation with which all modes of chromatography can be individually performed. While some good work has been done toward this end, no truly universal chromatograph exists today. However, one can imagine that if it did, it would include minimally two high pressure pumps for generating gradients or delivering a modifier to a primary mobile phase, a high-pressure valve injector, an oven or means for column temperature control, and a detector that could be operated in both universal or selective modes and handle mobile phases ranging from gases to liquids. Figure 1 shows a general schematic of such a system. In this figure, the detector is depicted as a mass spectrometer (MS) with an atmospheric pressure ionization source, since this detector meets the detector requirements better than any other detector available today.

In order for a chromatographic instrument to be able to perform all types of column chromatography, there are a number of requirements that must be fulfilled. Since GC is most appropriately carried out using open tubular capillary columns, while LC requires a column packing material, it is reasonable to expect that the universal chromatograph would have to be designed around small columns of capillary dimensions. It was pointed out over 10 years ago that the practice of GC, SFC, and LC becomes more similar as the column diameter becomes smaller (2). Open tubular capillaries are typically used in GC, while packed capillary columns are becoming more popular for LC, especially when considering the coupling to MS. For SFC, either open tubular or packed capillary columns can be used, depending on the application.

For such a system, the mobile phase pumping system should be capable of delivering very low flow rates (μL min^{-1}) consistently and accurately under both isobaric or gradient conditions, and should be capable of both pressure and flow programming. A two-pump system would allow for modifier addition in SFC and SGC and binary mobile phase compositions for capillary LC. The chromatographic oven would have to be capable of temperature programmed operation as well as isothermal operation. An extremely critical concern for a capillary column-based system is the detailed attention that would have to be paid to the elimination of any dead volumes in the injector, connectors, and transfer lines. In addition, a pressure control device is required at the end of the column or after the detector to restrict the flow for SFC, or to provide atmospheric pressure conditions for GC, SGC, and LC.

Assuming that universal instrumentation could be constructed as described above, one would merely have to change the column and mobile phase, and make a few other adjustments, such as adding a restrictor, to change from one form of chromatography to another. Obviously, truly universal chromatography would not include a change in column when switching from one form of chromatography to another. The characteristics of universal chromatography and the properties of a truly universal column are treated in later sections.

Combined Chromatography

While a completely universal chromatographic system has never been constructed, a number of chromatographic systems have been combined with the objective of performing multiple chromatographic operations. The most common configuration of this type is the multidimensional arrangement, in which two or more columns are connected in series, and fractions from the first separation system are selectively transferred to one or more secondary separating systems to enhance resolution and sensitivity. Typically, two columns, each containing a different stationary phase, are combined in such a way as to perform GC-GC, LC-LC, SFC-SFC, LC-GC, LC-SFC, or SFC-GC. In most cases, two chromatographic systems are connected together with an appropriate interface between the columns in each system. Figure 2 shows a schematic of a microcolumn SFC-SFC system that used a valve-switching interface to heart-cut fractions from the first column, and send them to a second column for high resolution (3). Figure 3 shows a separation of a coal tar according to the number of aromatic rings in the first dimension, and then the resolution of two

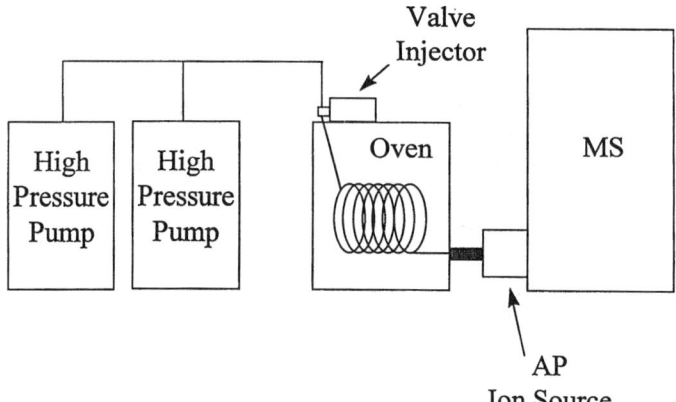

Figure 1. Schematic of a universal chromatographic system.

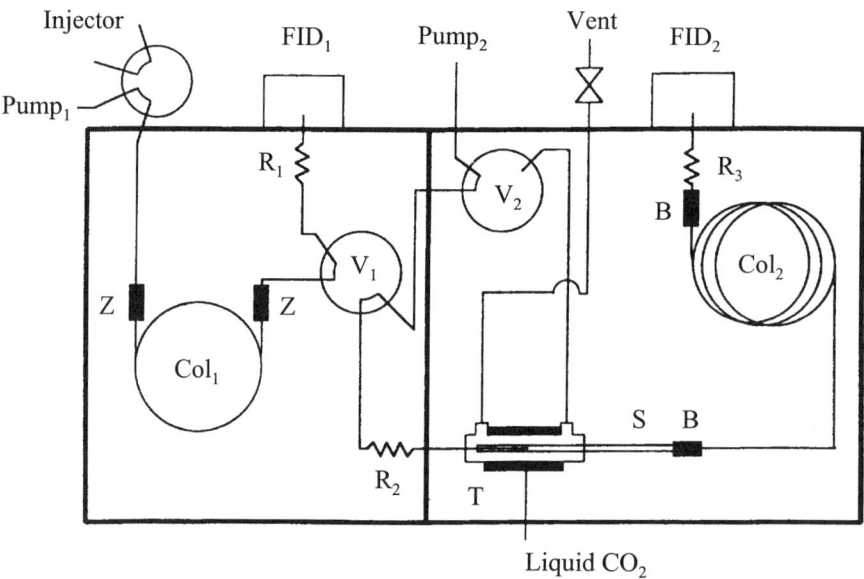

Figure 2. Schematic of a multidimensional capillary SFC-SFC system. Abbreviations: Col_1 = primary packed capillary column, Col_2 = secondary open tubular column, V_1 and V_2 = rotary valves, T = cold trap, S = capillary solute concentrator, R_1 and R_3 = frit restrictors, R_2 = linear restrictor, Z = zero dead volume union, B = butt connector. (Reproduced with permission from reference 3.) Copyright 1990 American Chemical Society.

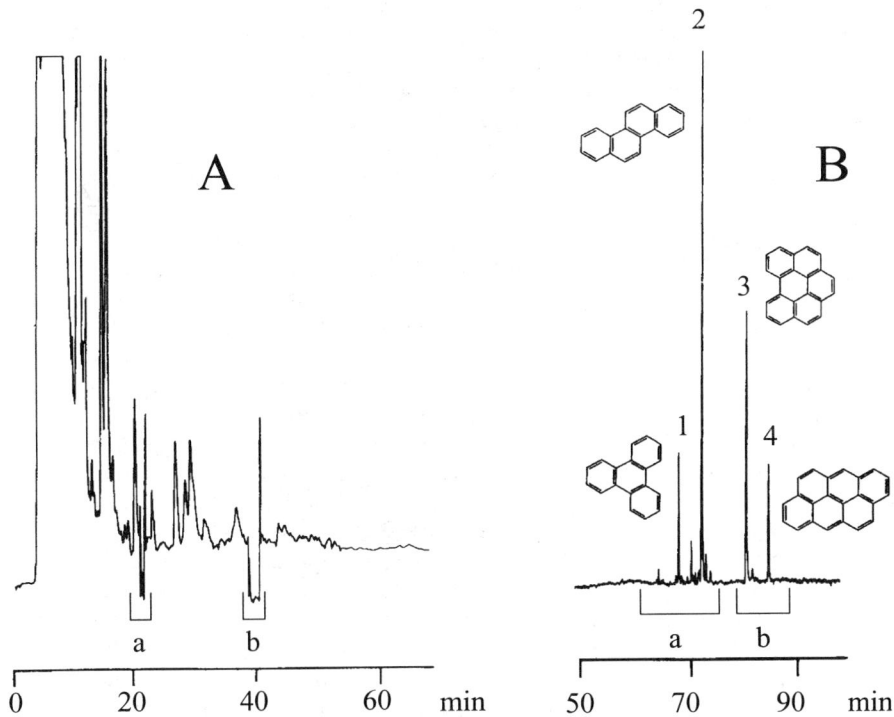

Figure 3. Separation of standard coal tar using a two-dimensional capillary SFC-SFC system. Two fractions were collected from the first dimension and then analyzed simultaneously in the second dimension. Cut "a" was collected between 20.2 and 21.2 min, and cut "b" between 38.7 and 40.2 min. Conditions: (A) carbon dioxide mobile phase, 75 °C, pressure program from 200 to 414 atm at 5 atm min^{-1}, (B) carbon dioxide mobile phase, 110 °C, pressure program from 70 to 120 atm at 20 atm min^{-1}, and then 120 to 414 atm at 8 atm min^{-1}, after a 2-min isobaric period, FID detection. Peak identifications: (1) triphenylene, (2) chrysene, (3) benzo[ghi]-perylene, (4) anthanthrene. (Reproduced with permission from reference 3.) (Copyright 1990 American Chemical Society.)

heart-cut fractions into single isomers according to shape in the second dimension. In this manner, high resolution separation of structural isomers can be obtained. Two-dimensional separations that involve heart-cutting are typically time-consuming because the separations occur in series.

Another limitation of typical two-dimensional separations is that most of the sample from the first column is not allowed to pass into the second column for further analysis. A more recent form of two-dimensional analysis, called "comprehensive" two-dimensional analysis, differs from the typical two-dimensional separations in that the total sample from the first dimension is allowed to enter the second dimension for further separation. In order for this to be possible, the second dimension must be much faster than the first dimension. Figure 4 shows a schematic of a comprehensive SFC-GC system in which the effluent from a capillary SFC column was introduced through a thermal desorption modulator into a high speed (25 μm i.d.) capillary GC column (*4*). Figure 5 shows a separation of polycyclic aromatic hydrocarbons using this two-dimensional system. As the 12.5-min SFC separation progressed, rapid 10-s GC separations were conducted. This two-dimensional chromatographic system was unique in that both columns were contained in the same oven, bringing multidimensional separations closer to the universal case.

SFC and GC can be combined in still another unique manner. Using a single chromatographic oven and column, GC can be performed by temperature programming the column with helium carrier gas, followed by switching to carbon dioxide and programming the pressure for SFC. Figure 6 shows a separation of a mixture of hydrocarbons and polydimethylsiloxanes in which the hydrocarbons were well-resolved by GC, however, the polymethylsiloxanes required the solvating power of SFC for elution (*5*). Using a more recent and improved system, either GC, SFC, LC, or sequential GC-LC could be performed by positioning a mode selector valve to introduce the appropriate mobile phase and select the appropriate detector (*6*). In the GC mode, the valve was positioned to introduce helium as the carrier gas and to direct the effluent to the flame ionization detector. In the SFC mode, the valve was positioned to select carbon dioxide and, again, the flame ionization detector. For LC, any of a variety of liquid mobile phases could be introduced and a UV-absorbance detector selected. Figure 7 shows a chromatogram of household wax obtained by sequential GC-SFC. Helium mobile phase was first used to resolve the volatile terpenes by GC with temperature programming, and then carbon dioxide was introduced to elute and separate the remaining compounds by pressure programming.

Unified Chromatography

In 1988, Ishii and Takeuchi introduced the terminology "unified chromatography" to describe the situation in which the different forms of chromatography, including GC, SFC, and LC, could be carried out with a single chromatographic system only

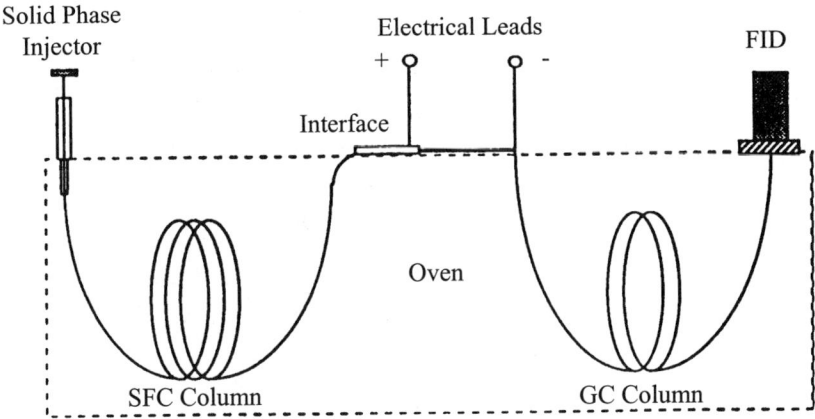

Figure 4. Schematic of a comprehensive SFC-GC system. (Reproduced with permission from reference 4.) (Copyright 1993 Vieweg Publishing.)

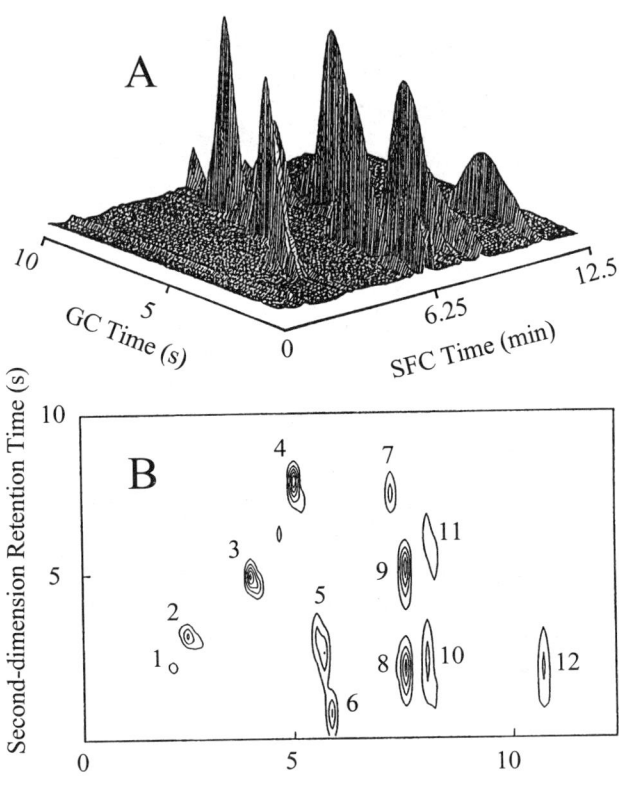

Figure 5. Separation of polycyclic aromatic hydrocarbons (PAHs) using a comprehensive SFC-GC system. Conditions: carbon dioxide mobile phase, temperature program from 125 to 200 °C at 8 °C min^{-1}, pressure program from 90 to 180 atm at 7 atm min^{-1}, FID detection. Thermal desorption modulator control parameters: 10 s time interval between secondary chromatograms, 100 Hz data frequency, 24 VDC by 20 ms duration modulation signal, 33-milliamperes per 10-s base current increment. Peak identifications: (1) naphthalene, (2) 2-methylnaphthalene, (3) acenaphthene, (4) fluorene, (5) 9-methylfluorene, (6) 1-methylfluorene, (8) phenanthrene, (9) anthracene, (10) 1-methylanthracene, (11) 2-methylanthracene, (12) pyrene. (Reproduced with permission from reference 4.) (Copyright 1993 Vieweg Publishing.)

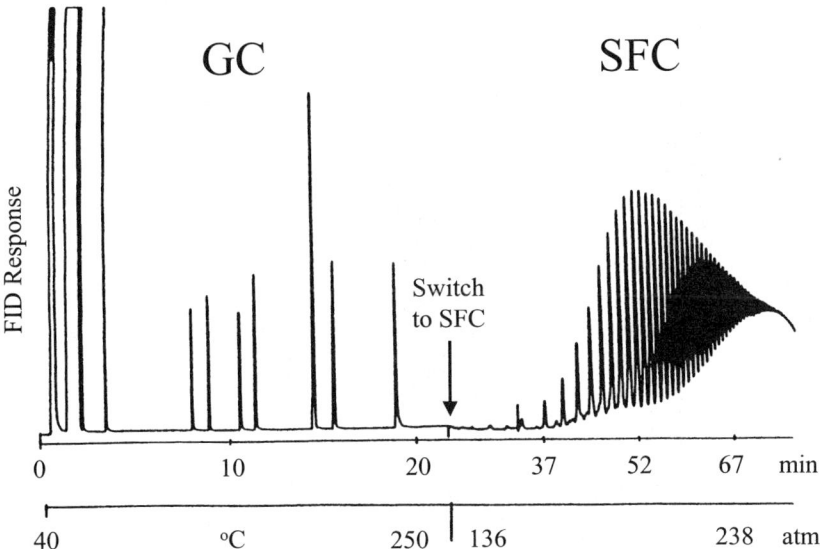

Figure 6. Separation of aromatic compounds and polydimethylsiloxanes using a combination of GC and SFC by switching the mobile phases during the same run. Conditions: 19.5 m × 100 μm i.d. fused-silica capillary coated with 0.4 μm film of DB-5; (GC) helium mobile phase, temperature program from 40 to 250 °C at 10 °C min^{-1}; (SFC) carbon dioxide mobile phase, pressure program from 136 to 238 atm at 2 atm min^{-1}, FID detection. (Reproduced with permission from reference 5.) (Copyright 1987 Preston Publications.)

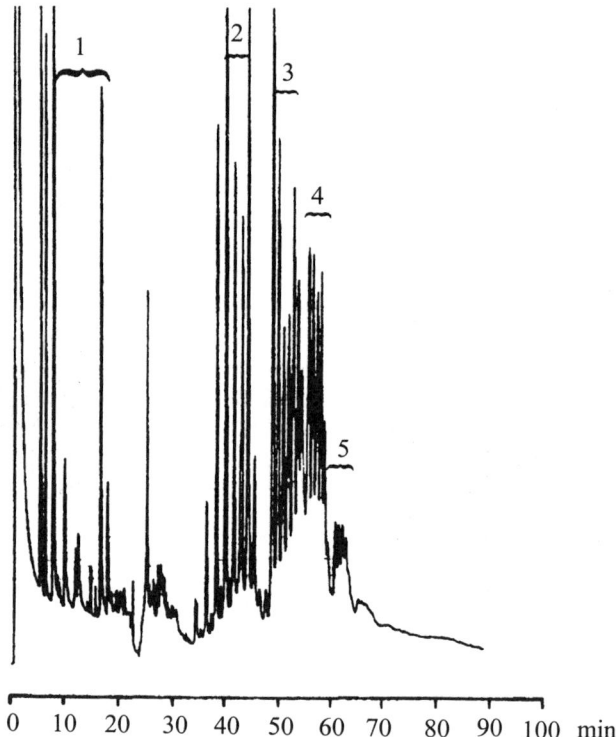

Figure 7. Separation of a dichloromethane solution of a household was using sequential GC-SFC. Conditions: 10 m × 50 μm i.d. open tubular fused silica capillary coated with SB-Methyl (d_f = 0.25 μm), valve injection, FID detection; (GC) helium mobile phase, 15 atm, 40–130 °C (5 °C min^{-1}); (SFC) carbon dioxide mobile phase, 120 °C, 120–360 atm (5 atm min^{-1}). Peak identifications: (1) terpenes; (2) C_{23}-C_{35} alkanes; (3) $C_{15}H_{31}$ COOR esters, R = $C_{24}H_{49}$ to $C_{34}H_{69}$; (4) diesters $C_{15}H_{31}$ COOC$_{15}H_{30}$ COOR, R = $C_{24}H_{49}$ to $C_{34}H_{69}$; (5) triesters. (Reproduced with permission from reference 6.) (Copyright 1997 Royal Society of Chemistry.)

by changing the column temperature and pressure (7). Using either an open tubular or packed capillary column and a mobile phase such as methanol or diethyl ether, they demonstrated wide differences in solute retention, depending on the temperature and pressure of the mobile phase. Furthermore, different modes of chromatography could be carried out in series in a single chromatographic run by merely adjusting the temperature and/or pressure. A mixture of aromatic hydrocarbons and styrene oligomers was separated using diethyl ether as mobile phase at 220 °C (diethyl ether has a critical temperature of 194 °C and a critical pressure of 35.6 atm), and programming the pressure from 24.2 atm. In this way, the hydrocarbons were separated in the GC mode prior to SFC separation of the styrene oligomers.

Universal Chromatography

"Universal chromatography" as defined in this chapter is almost synonymous with the terminology "unified chromatography," the only difference being the inclusion of the possibility that different modes of chromatography can occur simultaneously in a column during a run, i.e., any combination of gaseous, supercritical fluid, or liquid mobile phases can exist simultaneously at different locations in a column during an analysis. For example, Figure 8 shows a chromatogram of a mixture of aldehydes obtained in less than 1 min using a packed capillary column and carbon dioxide as the mobile phase where the mobile phase is introduced as a supercritical fluid at the column inlet and exits as a gas (8). Note that there is no phase transition occurring during this separation. This form of chromatography falls most appropriately in the classification of GC because there was no restrictor at the end of the column, and the mobile phase was a gas for some length of the column from the outlet. However, in order to obtain the rapid analysis shown in the figure, a high inlet pressure was used. In fact the pressure was programmed from 80 to 200 atm at 100 atm min^{-1}, which was always above the critical pressure of carbon dioxide. Therefore, the front of the column was under supercritical fluid conditions, the end of the column was under gaseous conditions, and there was a change from supercritical fluid to gas at some point along the length of the column. It should be noted that the occurrence of this change within the column had little, if any, effect on the chromatographic performance. A highly efficient separation was obtained.

The phase diagram in Figure 9 helps to visualize the effect of pressure and temperature on the properties of the mobile phase throughout the length of the column when using a packed column with a significant pressure drop. In Figure 8, the mobile phase at the front of the column was maintained above the critical pressure, and the whole column was maintained at a temperature above the critical temperature. Therefore, along with the pressure drop in the column, the mobile phase changed from a supercritical fluid into a gas as illustrated by arrow A in Figure 9. The same thing occurred (9) during the chromatography shown in Figure 10B. For comparison, when the column temperature was reduced to a value below

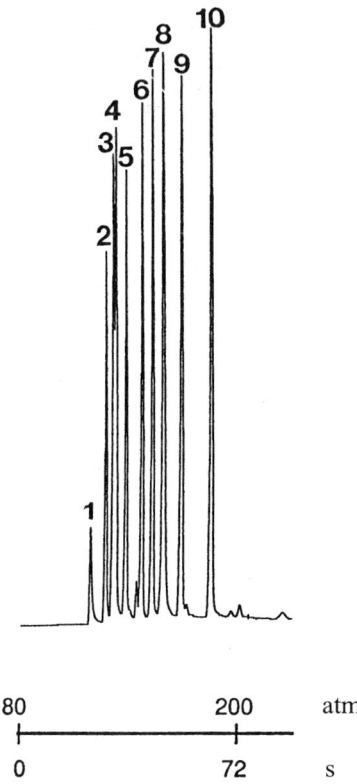

Figure 8. Separation of aldehydes using carbon dioxide at an inlet pressure and temperature above the critical point and the column outlet open to the atmosphere. Conditions: 30 cm × 250 μm i.d. fused-silica capillary column packed with 5 μm (120 Å) silica particles bonded with diol, coated with polyethylenimine (PEI) and end-capped with hexamethyldisilazane (HMDS), carbon dioxide mobile phase, 200 °C, pressure program from 80 to 200 atm at 100 atm min^{-1}, FID detection. Peak identifications: (1) acetaldehyde, (2) propionaldehyde, (3) acrolein, (4) isobutyraldehyde, (5) butyraldehyde, (6) isovaleraldehyde, (7) valeraldehyde, (8) crotonaldehyde, (9) caproicaldehyde, (10) heptaldehyde. (Reproduced with permission from reference 8.)

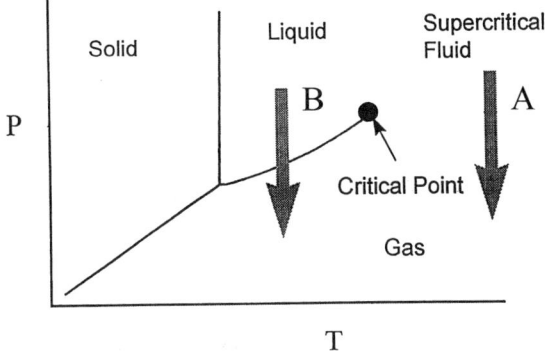

Figure 9. Phase diagram with arrows representing mobile phase transitions that occur when the column outlet is open to atmosphere and the inlet conditions are: (A) liquid mobile phase heated above is atmospheric pressure boiling point, and (B) supercritical fluid.

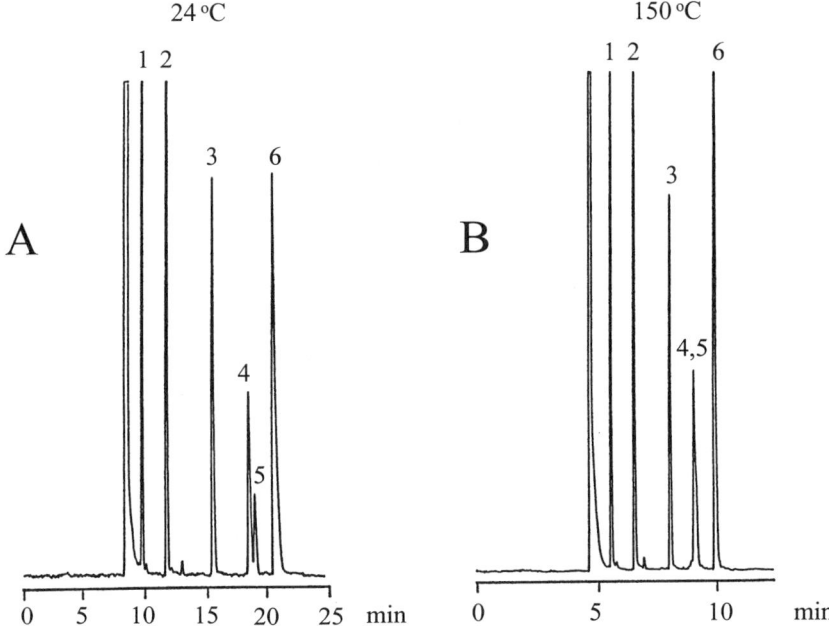

Figure 10. Separation of polycyclic aromatic hydrocarbons at column temperature (A) below and (B) above the critical temperature of the carbon dioxide mobile phase. Conditions: 2.2 m × 250 μm i.d. fused-silica capillary column packed with 10 μm silica (80 Å pores), 150 atm column inlet pressure, FID detection. Peak identifications: (1) n-hexane, (2) 1-hexene, (3) 1,5-hexadiene, (4) trans-1,3,5-hexatriene, (6) benzene. (Reproduced with permission from reference 9.) (Copyright 1999 Wiley.)

the critical temperature of the mobile phase (i.e., 24 °C), the carbon dioxide mobile phase was a liquid at the front of the column, and changed to a gas at some point along the length of the column. In this case, there was a distinct phase transition in the column (arrow B in Figure 9). Again, no disturbance in chromatographic performance was observed (Figure 10A). The lower temperature improved the selectivity for the *cis/trans*-hexatriene isomers, but also increased the analysis time.

Figure 11 shows a plot of reduced plate height *versus* linear velocity for the two conditions illustrated in Figure 10. These results verify that the phase transition from liquid to gas within the column had no observable effect on efficiency. The plots also illustrate the advantage of operating at higher temperatures when speed is desired. Figure 12 illustrates that similar results were observed when higher viscosity liquids (methanol, hexane, and acetonitrile) were introduced into the column at temperatures above their boiling points (*10*). Again, a transition occurred in the column from liquid to gaseous mobile phase conditions.

Obviously, if the temperature and/or pressure is changed during a chromatographic run, the conditions in the column can change dramatically. Figure 13 shows a chromatogram of a mixture of hydrocarbons obtained under liquid carbon dioxide mobile phase conditions initially at the front of the column, followed by increasing the pressure and programming the temperature until supercritical fluid conditions existed (*9*). Throughout the analysis, gaseous mobile phase conditions existed at the end of the column. At different times in this analysis, the mobile phase existed as a gas, liquid, and supercritical fluid, although never all three at the same time. The only way that all three could exist at the same time in the column would be to impose a temperature gradient along the length of the column, as well as a pressure gradient. There are two possibilities for this as is shown by the two dotted lines in Figure 14. In one case, the mobile phase would exist as liquid, supercritical fluid, and gas as it moved from the front to the end of the column, while in the other case, it would exist as a supercritical fluid, liquid, and then a gas.

Universal Column

In truly universal chromatography, it should be possible to perform all types of chromatography using a single column. It has already been suggested in this chapter that such a column must be a packed capillary column because of the slow diffusion associated with high density supercritical or liquid mobile phases. Packed capillary columns provide relatively high column efficiency per unit time, desirable retention characteristics, and good peak capacity. The challenge is to select the optimum column dimensions and particle size to maximize the performance in all of the separation modes. Theory clearly indicates that efficiency improves as the particle size is reduced. However, as the particle size decreases, the pressure drop along the column increases, which can have a deleterious effect on resolution when the mobile phase is compressible, such as is the case for a gas or supercritical fluid. This concern becomes more important when high speed separations are desired because

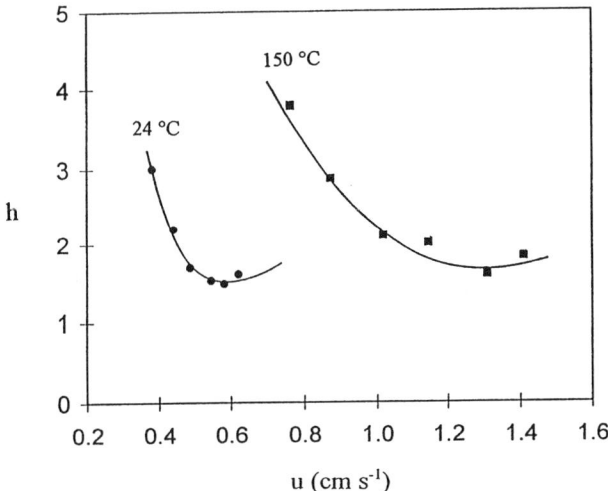

Figure 11. Plots of reduced plate height *versus* linear velocity at temperatures both above and below the critical point of carbon dioxide. Conditions: same as in Figure 10, hexane used as the test solute. (Reproduced with permission from reference 9.) (Copyright 1999 Wiley.)

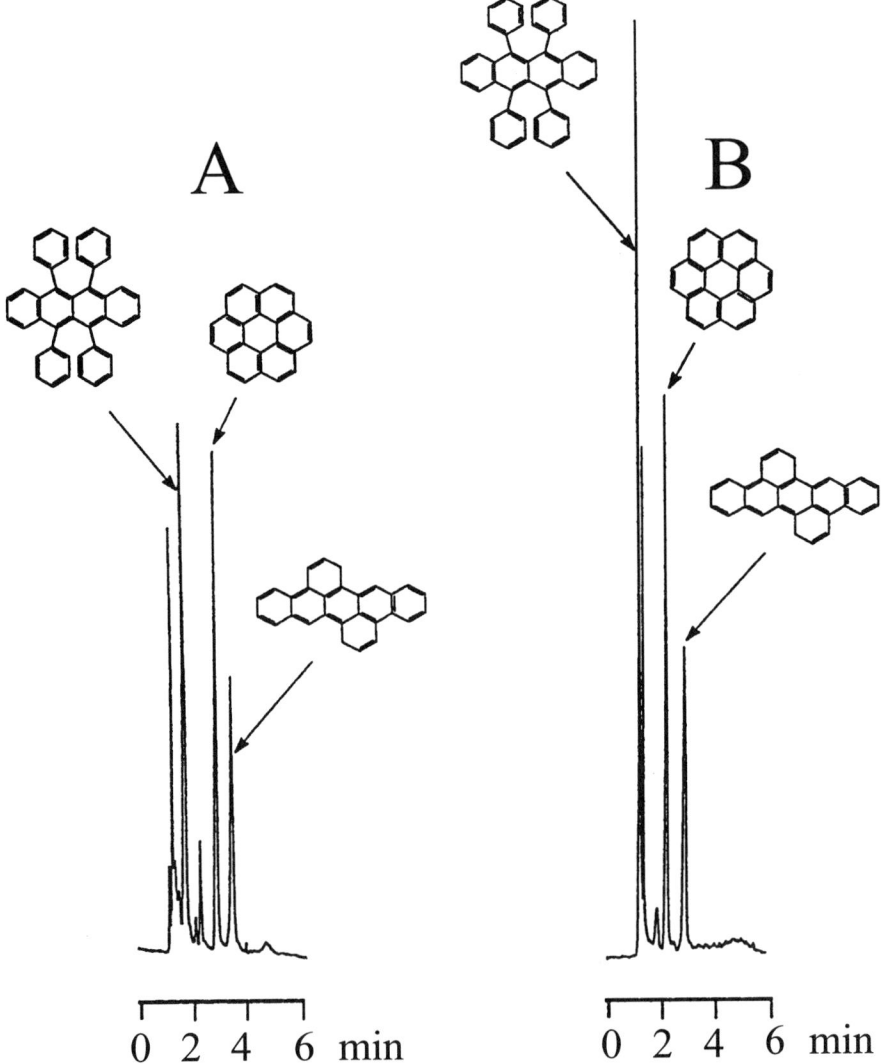

Figure 12. Separation of large polycylic aromatic hydrocarbons using mobile phases that transition from liquid to gas along the length of the coolumn. Conditions: 55 cm × 250 μm i.d. fused-silica capillary column packed with 5 μm polymeric ODS particles (200 Å pores), 300 atm column inlet pressure, temperature program from 65 to 250 °C at 25 °C min^{-1}, UV detection (254 nm). Mobile phase: (A) acetonitrile, (B) hexane, (C) methanol. (Reproduced with permission from reference 10.) (Copyright 1998 Royal Society of Chemistry.)

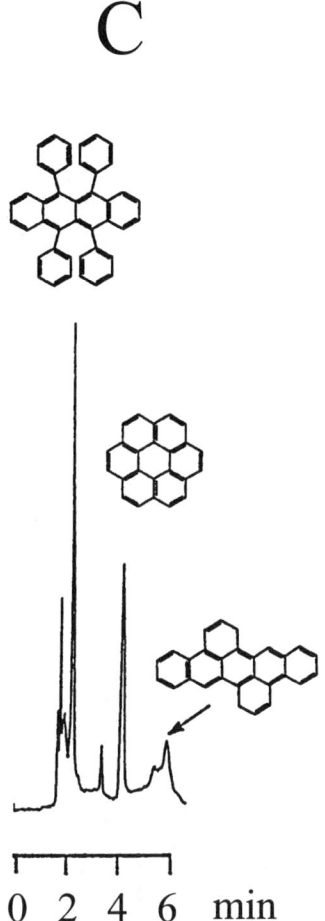

Figure 12. *Continued*

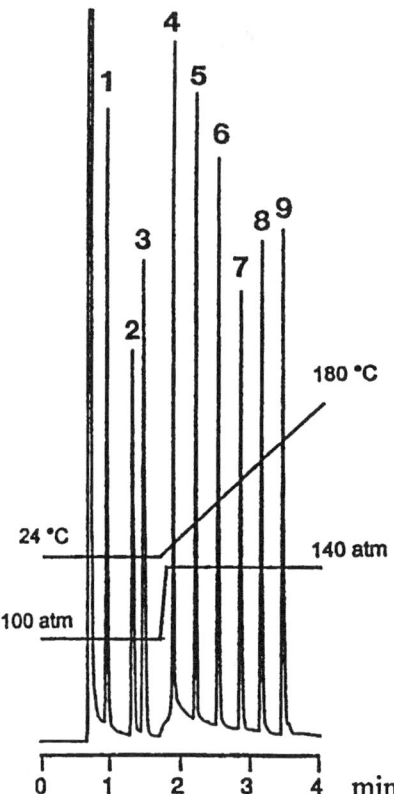

Figure 13. Separation of hydrocarbons using carbon dioxide mobile phase that transitions from liquid to gas and then from supercritical fluid to gas along the length of the column. Conditions: 45 cm × 250 μm i.d. fused-silica capillary column packed with 10 μm silica (60 Å pores), FID detection. Peak identifications: (1) n-hexane, (2) *trans*-2-hexene, (3) *cis*-2-hexene, (4) n-octane, (5) 1-octene, (6) n-decane, (7) 1-decene, (8) n-dodecane, (9) 1-dodecene. (Reproduced with permission from reference 9.) (Copyright 1999 Wiley.)

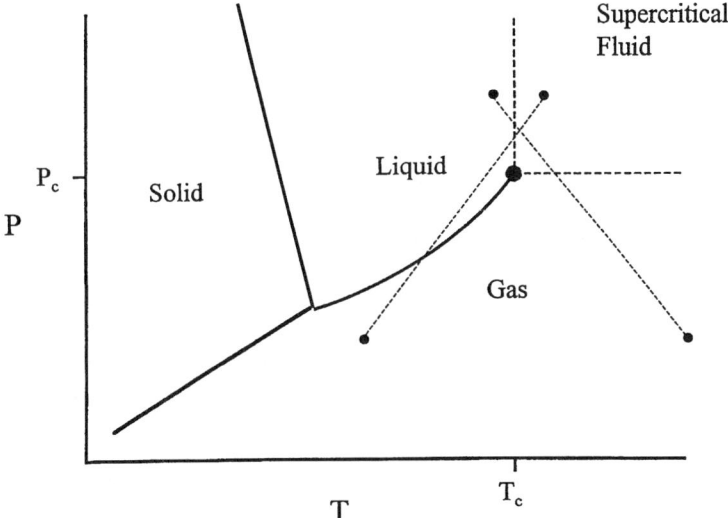

Figure 14. Phase diagram illustrating transitions which can occur under universal chromatographic conditions.

higher speed is a result of higher column head pressure, and higher column head pressure leads to higher pressure drop.

Chromatographic resolution depends on column efficiency (N), selectivity (a), and retention factor (k). Long analysis time can result in the highest resolution, however, for fast separations, the resolution that can be obtained per unit time (R_{st}) should be considered. For SFC, R_{st} can be represented by (11):

$$R_{st} = \frac{\sqrt{N_t}}{2\sqrt{t_M}} \frac{(a-1)}{a} \frac{k}{(1+k)^{1.5}} \qquad (1)$$

The value for R_{st} is dependent on the column efficiency per unit time (N_t), as well as on the retention factor. Figure 15 shows a plot of plates per second *versus* mobile phase linear velocity in cm s^{-1}. Notice that at high linear velocity, the maximum efficiency per unit time was obtained using 15 μm particles in the packed capillary. Columns packed with either 5 μm or 25 μm particles gave worse results. This is because the 5 μm particles produced a large pressure drop along the column, resulting in a large density drop and a significant increase in solute retention factor.

For fast separations, the resolution obtained per unit time is more important than the efficiency per unit time. Although columns packed with small particles provide high separation resolution, the increased retention caused by the density drop along the column leads to lower resolution per unit time. Figure 16 again shows that the 15 μm particles provide the best performance for high speed separations under SFC conditions. These theoretical predictions have been verified from actual chromatographic analysis (Figure 17). While analysis times for both separations were less than one minute, the column containing 15 μm particles gave a wider separation window and better resolution than the column containing 5 μm particles.

Other practical considerations, such as ease of interfacing to mass spectrometry, must be made in selecting the optimum universal column. Of course, smaller diameter columns produce flow rates that are easier to handle by mass spectrometer vacuum systems. With all of the various factors considered, the optimum universal column would be a packed capillary column (approximately 10-50 cm in length and 250 μm i.d.) packed with particles with diameters between approximately 7-15 μm.

Universal Column Efficiency Equation

In order to realize high speed separations in chromatography, high pressures must be applied to the mobile phase at the head of the column which, as previously discussed, leads to a compromise in chromatographic performance when compressible mobile phases are used. The compressibility of fluids affect three physicochemical factors (viscosity or fluidity, solute diffusivity, and solvating

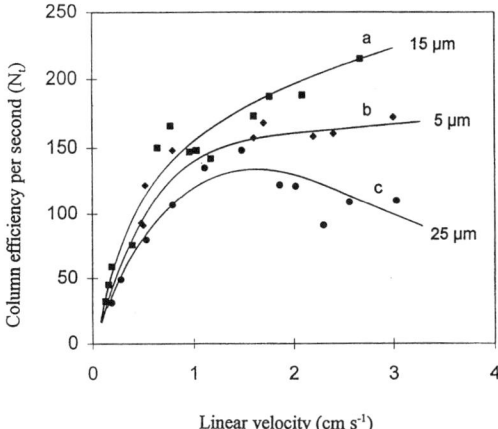

Figure 15. Plot of relationship between column efficiency per unit time and particle size in packed capillary SFC. Conditions: (a) 37 cm (b) 39 cm (c) 40 cm × 250 μm i.d. fused-silica capillary packed with various sizes of polymethylhydrosiloxane-deactivated and SE-54 encapsulated nonporous particles with film thicknesses of 0.1 μm, carbon dioxide mobile phase, 100 atm, 70 °C, FID detection, n-octane used as solute, methane used as an unretained marker. (Reproduced with permission from reference 11.) (Copyright 1997 Vieweg Publishing.)

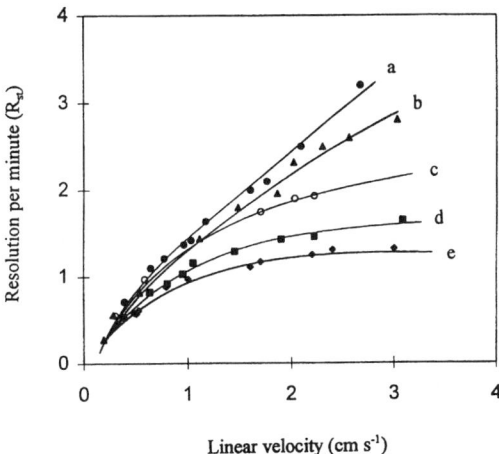

Figure 16. Plot of relationship between resolution per unit time, column length, and particle size in packed capillary SFC. Conditions: (a) 37 cm × 250 μm i.d. fused-silica capillary packed with 15 μm particles, (b) 40 cm × 250 μm i.d. fused-silica capillary packed with 25 μm particles, (c) 20 cm × 250 μm i.d. fused-silica capillary, (d) 30 cm × 250 μm i.d. fused-silica capillary, and (e) 39 cm × 250 μm i.d. fused-silica capillary packed with 5 μm nonporous particles, carbon dioxide mobile phase, 70 °C, 100 atm inlet pressure, n-octane as test solute, methane used as an unretained marker, FID detection. (Reproduced with permission from reference 11.) (Copyright 1997 Vieweg Publishing.)

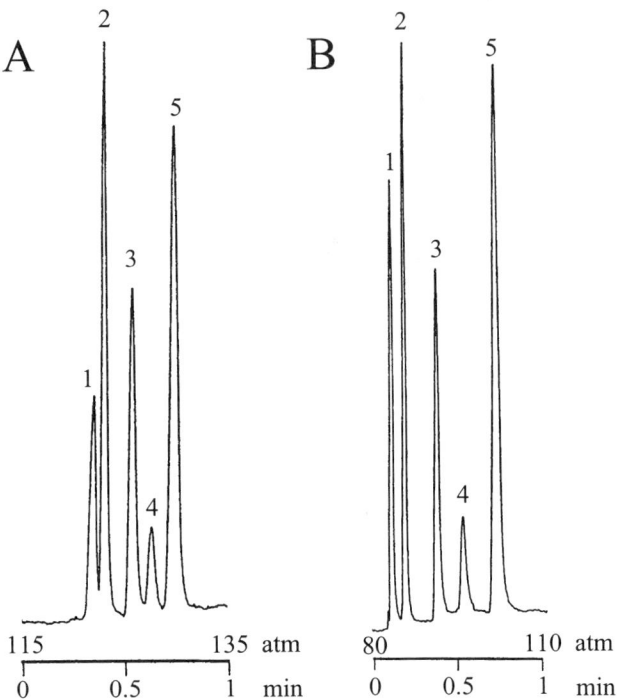

Figure 17. Separation of hydrocarbons using high speed packed capillary SFC. Conditions: (A) 9 cm × 250 μm i.d. fused-silica capillary packed with 5 μm and (B) 37 cm × 250 μm i.d. fused silica capillary packed with 15 μm polymethylhydrosiloxane-deactivated and SE-54 encapsulated nonporous silica particles with a film thickness of 0.1 μm, carbon dioxide mobile phase, 70 °, FID detection. Peak identifications: (1) carbon disulfide, (2) octane, (3) decane, (4) undecane, (5) dodecane. (Reproduced with permission from reference 11.) (Copyright 1997 Vieweg Publishing.)

power), which in turn influence chromatographic performance. A number of treatments over the years have addressed the effect of mobile phase compressibility, especially for gas chromatography, on chromatographic efficiency. These treatments have generally been based on average parameters and not on the actual conditions along the length of the column. Martire and Poe gave a rigorous treatment using the mobile phase parameters along the column to describe the column efficiency (12). While this new approach gives an accurate description of the column efficiency for compressible mobile phases, any relationship to the simple and classical van Deemter equation is lost.

Recently, the van Deemter equation was modified to broaden its range of applicability to cover compressibility and, hence, diffusion and viscosity changes in the mobile phase along the length of the column (13). The corrections were designed to incorporate dynamic variables into the mobile phase linear velocity terms while keeping the coefficient in each term independent of the mobile phase velocity or column pressure. The result is as follows:

$$H = 2\lambda d_p + \frac{2\gamma P_0^n D_0 d_p^{2n}}{(b\phi L)^n \eta_0^n u^{n+1}} + \frac{f_m(k)(b\phi L)^n \eta_0^n d_p^{2-2n}}{P_0^n D_0} u^{n+1} \qquad (2)$$

where P_0, D_0, and η_0 are the pressure, diffusion coefficient, and viscosity of the mobile phase at a defined pressure, b is a constant, and $f_m(k)$ is the relationship of H to k as k changes along the column. The other parameters in the equation are as defined in the common chromatographic literature, i.e., λ is the structural factor, d_p is the particle diameter, γ is the tortuosity factor, ϕ is the column permeability, and u is the average mobile phase linear velocity. The simplified van Deemter equation then becomes:

$$H = A + B/u^{n+1} + Cu^{n+1} \qquad (3)$$

where n varies between 0 and 1. For gaseous mobile phases, $n = 1$; for supercritical mobile phases, $n = 0.4$-0.6; and for liquid mobile phases, $n = 0$.

Conclusions

Generally in the past, any suggestion of the possibility of unified or universal chromatography has been taken lightly, primarily because of the different requirements in handling gaseous and liquid mobile phases. The development of supercritical fluid chromatography has helped tremendously to unify the chromatographic methods, since supercritical fluids can behave like gases (low pressure) or more like liquids (high pressure), and SFC instrumentation is a hybrid, containing some components utilized in GC and others in LC. In recent years, interest in universal chromatography has grown significantly. A universal chromatograph would simplify the practical aspects of chromatography, allowing a broader range of applications to be handled by a single instrument, making method development easier, and simplifying the maintenance of instrumentation. Universal chromatography has been demonstrated to various degrees using modified equipment in the research laboratory. The question that remains is whether or not a universal chromatograph can perform the various modes of chromatography to the level desired without a serious compromise in performance. The best case for universality in chromatography lies in high speed separations for which high pressures are required and differences in mobile phases are minimal. In such cases, miniaturization of column dimensions is essential.

Literature Cited

1. Myers, M.N.; Giddings, J.C. *Sep. Sci.* **1966**, *1*, 761.
2. Yang, F.J.; *Microbore Column Chromatography: a Unified Approach to Chromatography*; Marcel Dekker: New York, NY, 1989.
3. Juvancz, Z.; Payne, K.M.; Markides, K.E.; Lee, M.L. *Anal. Chem.* **1990**, *62*, 1384.
4. Liu, Z.; Ostrovsky, I.; Farnsworth, P.B.; Lee, M.L. *Chromatographia* **1993**, *35*, 567.
5. Pentoney, Jr., S.L.; Giorgetti, A.; Griffiths, P.R. *J. Chromatogr. Sci.* **1987**, *25*, 93.
6. Tong, D.; Bartle, K.D.; Clifford, A.A.; Robinson, R.E. *Analyst* **1995**, *120*, 2461.
7. Ishii, D.; Takeuchi, T. *J. Chromatogr. Sci.* **1989**, *27*, 71.
8. Wu, N.; Shen, Y.; Lee, M.L. *J. High Res. Chromatogr.*, accepted.
9. Shen, Y.; Lee, M.L. *J. Microcol. Sep.*, **1999**, *11*, 359.
10. Shen, Y.; Lee, M.L. *Anal. Chem.* **1998**, *70*, 737.
11. Shen, Y.; Lee, M.L. *Chromatographia* **1997**, *45*, 67.
12. Poe, D.P.; Martire, D.E. *J. Chromatogr.* **1990**, *517*, 3.
13. Shen, Y.; Lee, M.L. submitted.

Chapter 12

Practical Advantages of Packed Column Supercritical Fluid Chromatography in Supporting Combinatorial Chemistry

T. A. Berger

Berger Instruments, 123A Sandy Drive, Newark, DE 19713

Packed column supercritical fluid chromatography (SFC) is a form of normal phase chromatography, which allows higher throughput and enhanced peak recovery, compared to reversed phase HPLC, in high throughput screening applications. The inherent speed and viscosity advantages of carbon dioxide based fluids over normal liquids are exploited to produce very steep composition gradients at high flow rates through screening columns. Sample cycle times as short as 1.875 minutes produced peak capacities approaching 30, and differentiated solutes with a range of partition coefficients > 100. Such an approach allowed the detailed analysis of up to 770 samples/day. Hardware requirements for generating such throughput are discussed. Use of analytical scale instrumentation to collect fractions in the multiple mg range is also demonstrated. Columns 4 mm in diameter were shown to be capable of separating at least 6.5 mg of favorable compounds.

As with any technique, a good deal of historical effort has been expended to differentiate supercritical fluid chromatography (SFC) from the more established alternative chromatographic techniques: gas chromatography (GC) and high performance liquid chromatography (HPLC). However, from its start more than 35 years ago, SFC has also been viewed as the most important bridging technology between GC and HPLC, and, more recently, has sometimes been used to show that the three techniques are actually parts of a continuum. This chapter on packed column SFC was created as part of an ACS symposium on "unified chromatography." The theme of this symposium is an attempt to break down the apparent borders between all these chromatographic techniques.

A user can start a chromatographic experiment, which can clearly be defined as GC. By increasing the pressure (and possibly temperature), the name for the experiment eventually changes to SFC. Subsequently, a modifier could be added. A decrease in temperature produces only a change in the definition of the technique to HPLC (with a compressible fluid). Some people call this region subcritical fluid chromatography. At still lower temperatures, many fluids (i.e., pentane) liquefy at atmospheric pressure, and any difference between SFC and HPLC vanishes. Throughout such an experiment, there was only one phase present. Some have defined an additional instrumental subset called "enhanced fluidity" chromatography, where a small amount of a gas is added to a liquid to decrease viscosity and increase diffusivity, compared to normal HPLC. There is extensive literature showing solvation characteristics of dense gases, proportional to pressure, below the critical pressure of the fluid (subcritical).

SFC has always been misnamed because its name suggests the solvent characteristics of interest are unique to supercritical conditions. All these techniques exploit the same fluid characteristics to a greater or lesser extent (i.e., super- and sub-critical fluid chromatography are identical. Enhanced fluidity chromatography is simply subcritical fluid chromatography at high (i.e., 90%) modifier concentrations). The trend toward using much hotter solvents in HPLC is another form of enhanced fluidity or subcritical fluid chromatography.

All these techniques use the same hardware in the virtually the same way. They all use the same form of packed columns, operated with flow control pumps (to get accurate flow and fluid composition). In the past, the biggest mechanical difference was the use of a back pressure regulator (BPR) in SFC, used to raise the outlet pressure above ambient. However, it is now also common to put a small BPR on the outlet of HPLC's to prevent outgassing of dissolved gases (two phase formation). The only hurdle that separates SFC from HPLC is the inadequacy of most HPLC hardware for pumping compressible fluids. Hardware for performing SFC is perfectly capable of performing HPLC, but it is almost never true that HPLC equipment can properly perform SFC.

The advantages of packed column SFC, particularly as it is used in the pharmaceutical industry, stem from practical characteristics (low viscosity, high diffusivity) of carbon dioxide based fluids, regardless of whether the fluid is defined as super- or sub-critical. Under most practical conditions, the technique called packed column SFC is actually an odd form of HPLC with a highly compressible solvent.

SFC in High Throughput Screening

Pharmaceutical companies performing combinatorial chemistry or massively parallel synthesis are generating tens of thousands of new compounds per month, most of which require HPLC confirmation of purity. At such high analysis volume, many instruments and many operators are required to try to keep up. The cost of purchasing, storage, and disposing of solvents becomes a problem.

The general approach for maximizing throughput is to use shorter columns, smaller particles, and high flow rates. A 15 cm column with 3μm particles is as efficient as a 25 cm column with 5μm particles but generates that efficiency in less than 40% of the time. Similarly, a 7.5 cm column with 1.5μm particles is as efficient but >10 times faster than the column with 5μm particles. Additional speed can also be achieved by increasing flow rates above the optimum value. Doubling the flow rate cuts efficiency approximately in half but only cuts resolution by the square root of 2.

In HPLC screening applications, 5 cm long columns with 3μm particles are commonly run at up to 3 ml/min. A more extreme approach uses 15-20 mm long guard columns at 5 ml/min.

Although many people have recognized that SFC is faster than HPLC, in more traditional applications, this advantage has often not seemed great enough to overcome the inertia of the better-established technique. Further, packed column SFC is a normal phase technique. Since the separation mechanism is substantially different from reversed phase HPLC (rHPLC), solute selectivity tends to be quite different. For those looking for the same selectivity in shorter time, this can be viewed as a problem. On the positive side, packed column SFC offers complimentary selectivity to reversed phase HPLC and/or capillary electrophoresis (CE).

Normal phase HPLC is seldom practiced today, due to numerous practical problems, such as slow equilibration times, difficulty running gradients, drifting retention times caused by traces of water in the solvents, and the generation of large volumes of flammable waste. These difficulties do no apply to SFC.

The diffusion coefficients of solutes in the mobile phase determine the optimum linear velocity (flow rate) and analysis time. At the same temperature, diffusion coefficients in carbon dioxide based fluids are as much as an order of magnitude higher than in the liquids commonly used in rHPLC. The optimum linear velocity of a chromatographic mobile phase is at least 3 to 5 times greater in packed column SFC than it is in HPLC (*1*), **as shown in Figure 1.** For any given particle size, the optimum flow rate in SFC is substantially higher than in HPLC, and the shape of the vanDeemter curves is much flatter. This flatness favors operation above optimum velocity, since little efficiency or resolution is lost with shorter analysis times.

The pressure drop across a column is directly proportional to the viscosity of the fluid. The lack of intermolecular interaction in carbon dioxide results in viscosity an order of magnitude lower (1/10th) than in normal liquids (*2*), allowing HPLC columns to be operated at higher flow rates (higher linear velocity), with lower pressure drops.

Pure carbon dioxide is a relatively non-polar solvent, similar to hexane or some fluorocarbons. Polar solutes are soluble to only ppm ($1/1 \times 10^{-6}$) (*3*) to ppb ($1/1 \times 10^{-9}$) (*4*) levels. Mixed with polar solvents, the solvent strength of the mixtures, and the solubility of polar solutes increases. Methanol is the most polar, common modifier completely miscible with carbon dioxide over a wide range of

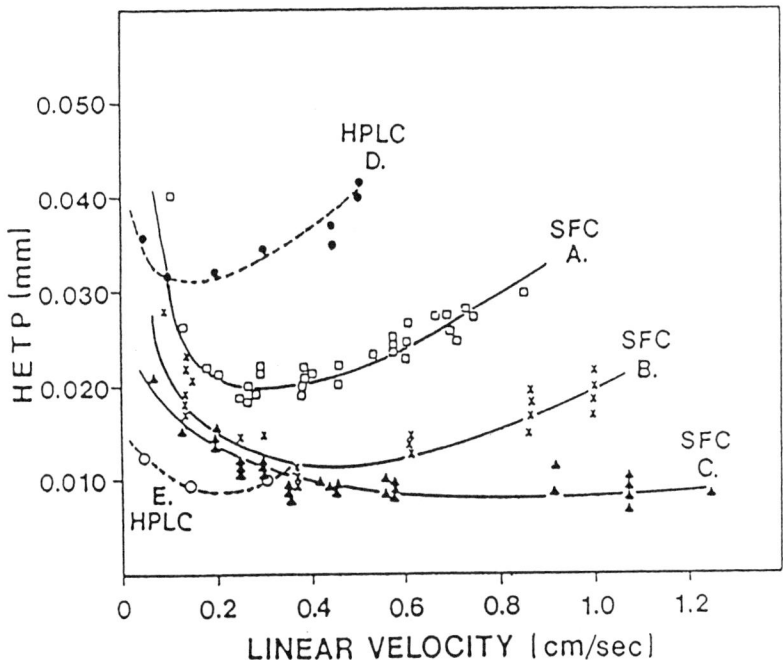

Figure 1. Experimental vanDeempter plots showing the relative speed of SFC and HPLC on the same columns. A and D = 10μm particles, B = 5μm, C and E = 3μm.

temperatures and pressures. Carbon dioxide/methanol mixtures cover a wider range of solvent strength than any pair of organic solvents used in HPLC.

With binary fluids, pressure becomes a secondary control parameter (*5*), with much less influence on retention or selectivity than mobile phase composition, as seen in **Figure 2**. As in HPLC, the primary mode for screening a wide range of solute polarity involves composition gradient programming.

A major advance in packed column SFC occurred with the introduction of a third more polar component into the mobile phase (*6-8*). Polar additives, such as trifluoroacetic acid or isobutylamine, allow the elution of virtually all small molecules used in pharmaceuticals. With the inclusion of additives into the mobile phase, virtually all members of families elute with similar peak shapes.

It should be remembered that there is no border between SFC and HPLC. In SFC screening methods, one could program from 0 to 100% modifier in a single run. However, at high modifier concentrations many of the practical advantages of SFC are lost, as the characteristics of the fluid (i.e., high diffusion coefficients and low viscosity) approach those of normal liquids (as the chromatography approaches standard HPLC (i.e., 100% modifier)). Surprisingly, little has been written about this transition from SFC to HPLC.

As a rule of thumb, doubling the modifier concentration halves retention (*9*). Solvent strength is a non-linear function of composition, as shown in **Figure 3**. The first small addition of methanol results in a larger than expected increase in solvent strength. This enhanced solvent strength is due to a well-known phenomenon called clustering (*10*). The polar modifier molecules tend to cluster together creating enhanced localized concentrations. Such clustering occurs in some mixtures of normal liquids but is much more pronounced in mixtures of carbon dioxide/methanol.

Little additional elution strength is achieved by using modifier concentrations much above 50 or 60% **(Figure 3)**. It is advantageous to avoid higher modifier concentrations to maximize efficiency per time and to minimize pressure drops. In screening methods, an additional hold time at the maximum concentration (i.e., 50-60%) allows elution of the last eluting components.

There is little incentive to set the temperature higher than 30 to 50°C. Higher temperatures are avoided to insure that thermally labile solutes are not degraded. The outlet pressure is set at moderate levels (i.e., > 100 bar). Under such initial conditions, the fluid is usually defined as supercritical. Upon increasing the modifier concentration, the critical temperature and pressure of the mixture are likely to increase. If the temperature is below the critical temperature of the new mixture, the definition of the fluid changes to a liquid. However, in virtually all practical circumstances, it is irrelevant whether these fluids are defined as subcritical or supercritical. No significant changes in either physical or chemical characteristics occur when the definition of the fluid changes.

The critical point of binary mixtures is often higher than either pure fluid. The literature sometimes suggests that the temperature and pressure should always be set higher than the maximum possible critical conditions of any combination of

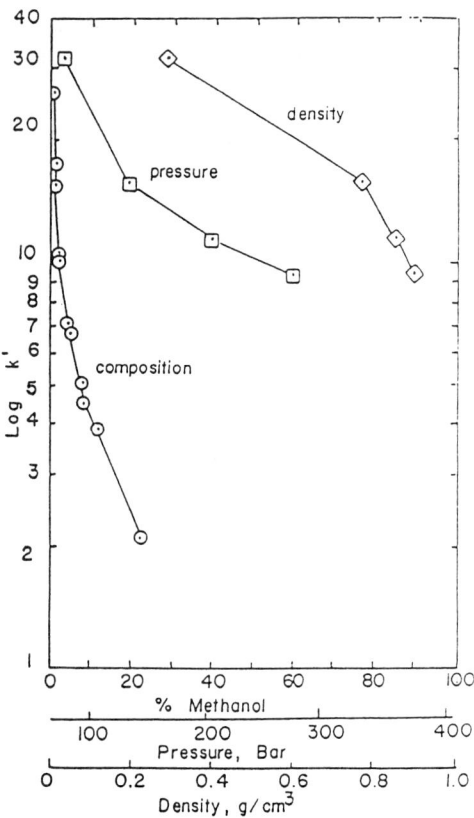

Figure 2. Composition is more important than density when using binary fluids.

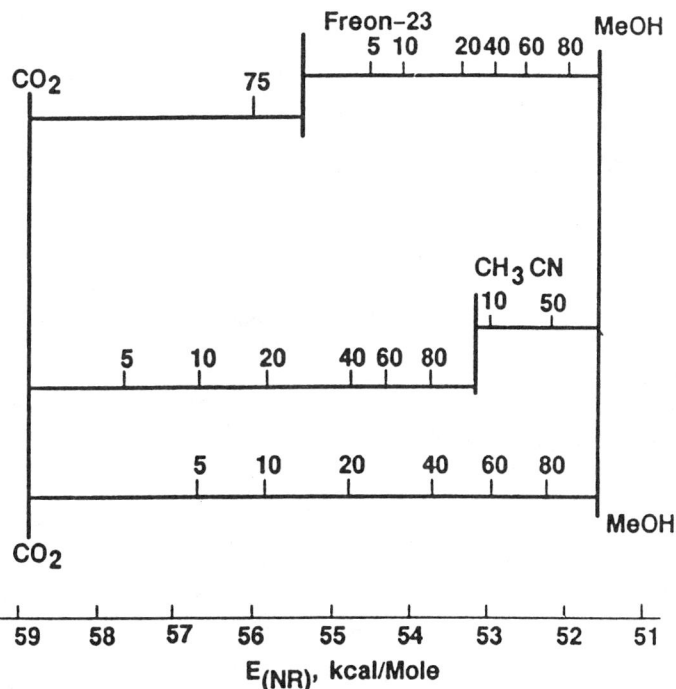

Figure 3. Solvent strength is a non-linear function of composition. Small numbers next to vertical lines indicate % modifier in carbon dioxide.

the two components to avoid the possibility of phase separation. Such an approach is excessively conservative.

The enclosed space at the bottom of **Figure 4** outlines a two-phase region for methanol in carbon dioxide (*11*) at one temperature. Compositions inside this region cannot be made. However, operating above the two-phase region is trivial, by simply setting the column outlet pressure above, in this case, 80 bar.

Once modifiers are added, both pressure and temperature become secondary control variables. As a practical matter, the column outlet pressure and column temperature is usually fixed. Elution strength is adjusted by changing the concentration of the mobile phase.

The diagonal lines in the upper part of **Figure 4** are constant density lines. Changing the composition at constant temperature and pressure causes small changes in density. However, the main reason for changes in retention is a change in the composition, not a change in density.

Chromatographic Purification of Samples

There is a strong desire to combine high speed screening with peak collection to generate 10-50 or 100-300 mg of pure compound at early stages of testing. With modern methods of synthesis, very large numbers of samples can be made in a short time, increasing the demand for throughput in the subsequent chromatographic cleanup step. Packed column SFC holds at least the potential to replace much of semi-preparative HPLC. The inherent advantages in diffusion coefficients/optimum flow rate and viscosity, exploited in analytical scale chromatography, offer similar enhancements on a larger scale. Throughput should increase 3 to 5 times.

Unlike HPLC, most of the fluids used in SFC inherently want to vaporize. After the backpressure regulator, the fluid expands roughly 500 times. Two phases typically form; a small volume of liquid phase, consisting mostly of polar modifier, and a large volume gaseous phase, consisting mostly of carbon dioxide, at low density (i.e., $0.002 g/cm^3$). Polar solutes tend to be non-volatile and remain in the more polar liquid phase.

Compared to HPLC, the recovered polar solutes will be contained in a much smaller volume of more easily removed organic solvent. Even at a worst case modifier concentration of 60%, at the pumps, only 50% of the total solvent volume will be collected as a liquid (as much as 10% of the modifier will vaporize with the carbon dioxide). The average peak will be collected in 20% of the total solvent volume, and the solvent will be non-aqueous. Thus, the volume of liquid containing each peak, and the time required to remove it, is dramatically reduced, compared to rHPLC.

Many organizations do not allow normal phase separations on a large scale due to safety consideration. If a large number of instruments are used to clean up samples, collectively, they can generate large volumes of mixed waste. If all this

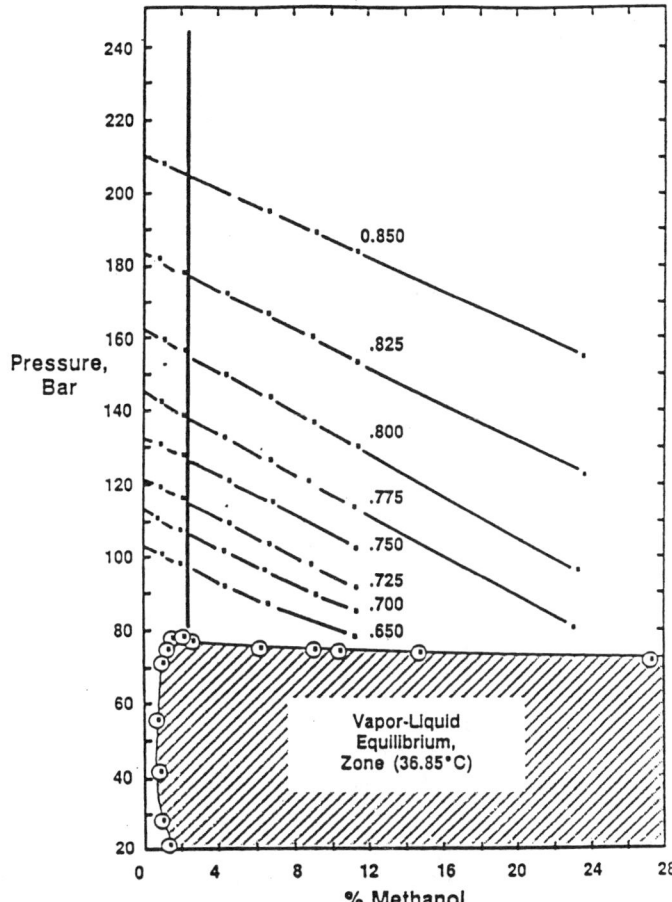

Figure 4. Density of carbon dioxide/methanol mixtures as a function of pressure and composition at 36.85°. Reproduced with permission from reference 11.

waste is organic and flammable, safe storage becomes an issue. In many locations, the cost of disposal now equals the original purchase price of the solvents.

If normal phase HPLC separations are ruled out, for safety reasons, the traditional alternative is reversed phase HPLC, with aqueous based mobile phases. From a practical perspective, removal of the aqueous based solvent from the compounds of interest becomes the limiting step in the use of the technique, being much slower than the chromatography. In fact, the use of semi-prep HPLC is often limited not by the chromatography, but by the ability to collect fractions, and subsequently the time required to evaporate off the (typically aqueous based) solvent.

Sample Solvent Capacity. A potential problem in SFC is the limited capacity of the columns to tolerate large volumes of polar sample solvents, such as methanol or DMSO. For 4 mm I.D. columns, injection volumes larger than approximately 10μl tend to cause peak broadening or distortion. There are techniques for solvent elimination, but they add to analysis time. A preliminary experiment shows that very concentrated samples can be injected and significant quantities can be separated and collected, even on a 4 or 10 mm column.

Experimental

A slightly modified SFC (Berger Instruments (BI), Newark, DE, USA) was used as the chromatograph. The system consisted of a FCM 1200 dual pump module, an ALS 3100 autosampler with a rack for microtiter plates, a TCM 2100 oven module, a Model 1050 diode array detector (Hewlett Packard (HP), Little Falls, DE, USA) with a high pressure flow cell, and a BI 3-D Chemstation.

The FCM 1200 was slightly modified by changing the volume of the mixing chamber. Several different mixing columns were used in these experiments. The best compromise between baseline noise and short gradient delay time was obtained using a mixing column/pulse dampener combination with a hold-up time of 6 sec. A more detailed description is presented in the discussion section.

For high speed separations, the autosampler performed full loop 10 μl injections, with methanol as wash solvent. Microtiter plates were received dry and reconstituted by adding methanol or dimethylsulfoxide (DMSO) to each well. Both shallow well and 4X wells were used. The autosampler was set to eliminate delays between injections. For some of the peak collection work, larger loops, up to 210μl were used. The microtiter plate rack was replaced with a rack for 2 ml vials.

The columns used for the bulk of the work all had a cyanopropyl (CN) bonded stationary phase. Some other experiments, using bare silica, diol, and amino phases, indicated that universal methods were more difficult (but not impossible) to achieve on these more polar phases than on the less specific, less retentive, cyano phase.

Columns used extensively included: 4 x 125 mm, 5 µm Lichrospher CN in a Lichrocart format, purchased from HP; Zorbax 50 x 4.6 mm, 3.5µm SB CN, donated by HP. Others included: 50 x 4.6 mm, 3µm Hypersil Silica (HP), 50 x 3 mm, 3µm Deltabond CN, provided by Keystone Scientific, Bellefonte, PA.

High Speed Gradient Conditions. The mobile phase was a linear gradient, typically 5-65% methanol containing either 0.5% isopropylamine (IPAm) or 0.4% trifluoroacetic acid (TFA). All chemicals used in the mobile phase were reagent grade or better. SFC grade carbon dioxide was purchased from Scott Specialty Gases, Plumsteadville, PA in both Berger Instruments Technimate cylinders or in larger steel cylinders, all with dip tubes but no headspace pad. The gradient rate was up to 45%/min. Total flow was 3 to 5 mls/min. Column temperature was 35°C. Outlet pressure was maintained at 120 bar.

Peak Collection. For sample recovery experiments, the SFC was modified as shown in **Figure 5.** The vapor pressure of methanol is high enough so that there is normally no liquid phase present after expansion until the methanol concentration exceeds approximately 10%. Any nonvolatile solutes emerging during the part of the gradient below 10-15% modifier will likely be trapped in the transfer line between the backpressure regulator and the fraction collector switching valve. The presence of a second liquid phase, which washes the lines and collects nonvolatile solutes, can be guaranteed by introducing a small additional liquid flow after the chromatography is complete.

An additional HPLC pump was added to the configuration. Its flow could be programmed to add fluid whenever the modifier concentration was below 10%. The flow could be turned off when the modifier concentration exceeded 15%.

The pump output was delivered through a tee containing a check valve mounted in the flow path after the UV detector. The check valve prevents backflow toward the make-up pump when its flow is shut off. This pump insures that there is always a liquid phase present to wash the transfer lines downstream of the backpressure regulator.

The optimum internal diameter of the transfer line is somewhat dependent on the flow rate. If the inner diameter is too small, aerosols tend to be formed. Increasing the modifier concentration tends to decrease aerosol formation. Nevertheless, in SFC it is a good idea to mount the fraction collector in a hood.

A device, similar to a miniature momentum separator was used to separate the gaseous phase from the liquid phase. This device is shown schematically in **Figure 6.** This device only partially solves the aerosol formation problems associated with the up to 500 fold expansion occurring across the back pressure regulator. Under some conditions aerosols escape out the top of the device, causing sample loss, and posing a health risk. Packing with glass wool to try to break the aerosol causes carry over problems. This device should NOT be used in an open laboratory, but confined to use in a fume hood.

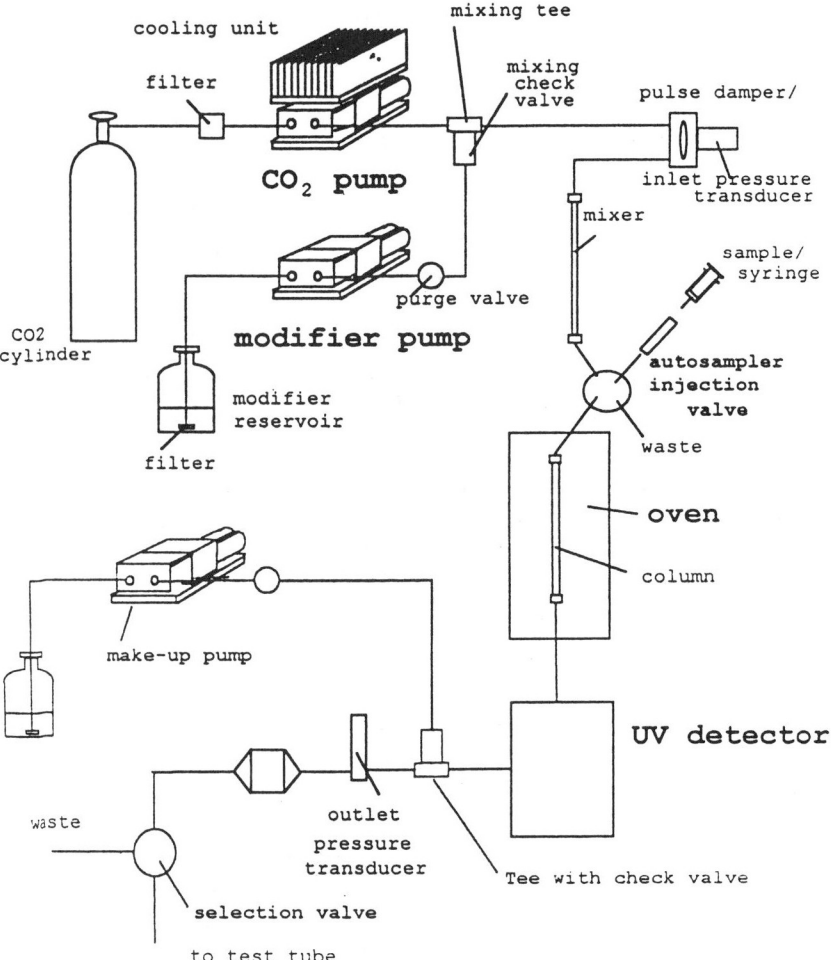

Figure 5. Schematic diagram of SFC modified to collect samples.

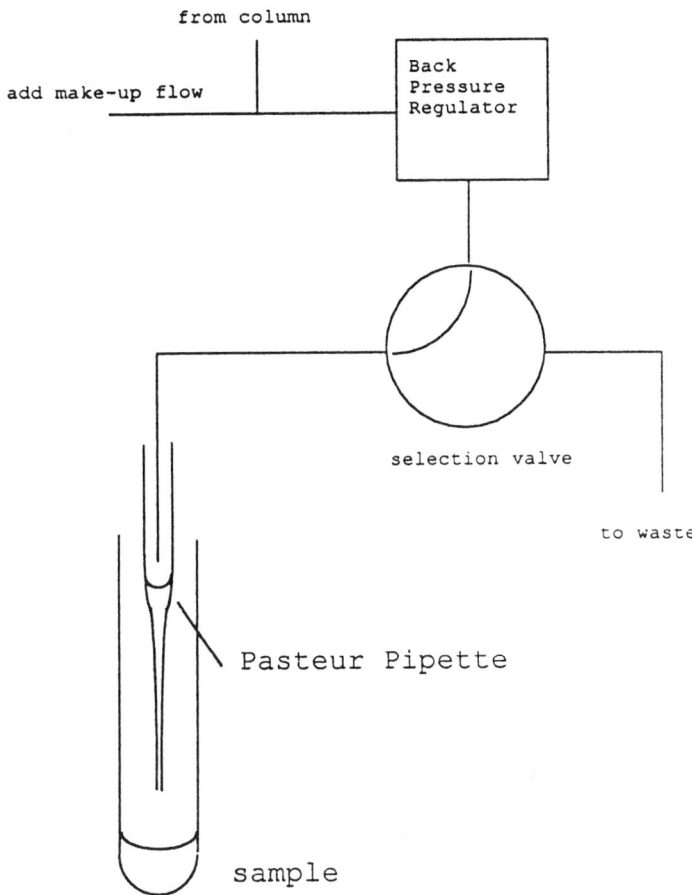

Figure 6. Schematic of miniature momentum separator used to collect fractions.

Results

High Speed Separations. A 50 mm x 4.6 mm column packed with 3.5 μm particles allowed rapid separations of a wide range of substituted amides, with a cycle time under 3 minutes. This corresponds to approximately 540 well/day, or 5.6-96 well microtiter plates/day. The gradient was 5-65% at 33%/min, with short initial and final hold times. Column temperature was set at 35°C. The flow rate was **4 mls/min**. The outlet pressure was maintained at 120 bar throughout. A series of chromatograms from consecutive wells on a microtiter plate are presented in **Figure 7**.

At the initial modifier concentration of 5% methanol, the column pressure drop was 9 bar (inlet pressure = 129 bar). During the course of the programming to 65%, the pressure drop increased dramatically to 270 bar (inlet pressure 390 bar), as the fluid characteristics approached those of a normal liquid (i.e., 100% methanol). A plot of inlet and outlet pressure vs. % modifier is shown in **Figure 8**. The pressure drop appears to be linearly related to % modifier, implying, at least roughly, that the viscosity of the mixture is linearly related to % modifier. The low pressure drop at low modifier concentrations, and the limitations arising at high concentrations is a direct demonstration of the practical advantages of SFC.

The chromatograms yield results similar to standard HPLC screening (number of peaks, peak capacity) but in approximately 1/3rd the analysis time. SFC is a normal phase technique. Elution order is typically inverted compared to reversed phase HPLC.

This SFC application produces approximately 1/2 the total liquid waste but with 3 times more samples run (1/6th the waste per run). The per liter cost of carbon dioxide and methanol tends to be low, compared to the cost of toxic solvents, such as acetonitrile. Waste disposal costs are also becoming an issue, particularly when many instruments are run in parallel. Significantly decreasing the amount of waste, thus, has a noticeable impact on total cost.

An alternate approach with a less aggressive column actually produced higher throughput, as shown in **Figure 9**. A 4 x 125 mm column packed with 5 μm particles operated at 5 mls/min produced much less pressure drop at higher flow rates. Pressure drop varied from 30 to 160 bar (inlet pressure 150 to 280 bar). By using a slightly steeper gradient (45%/min), a slightly higher initial (15%) and lower final (60%) modifier concentration, the cycle time (including autosampler access time and column re-equilibration time) was decreased to 1.875 minutes between injections. The actual run time was approximately 1.3 minutes. This corresponds to nearly 770 samples/day, or 8-96 well plates/day.

Peak Collection

Large Volume Injections. Progressively larger volumes of a complex mixture of an experimental substance dissolved in methanol were injected onto a 4 mm

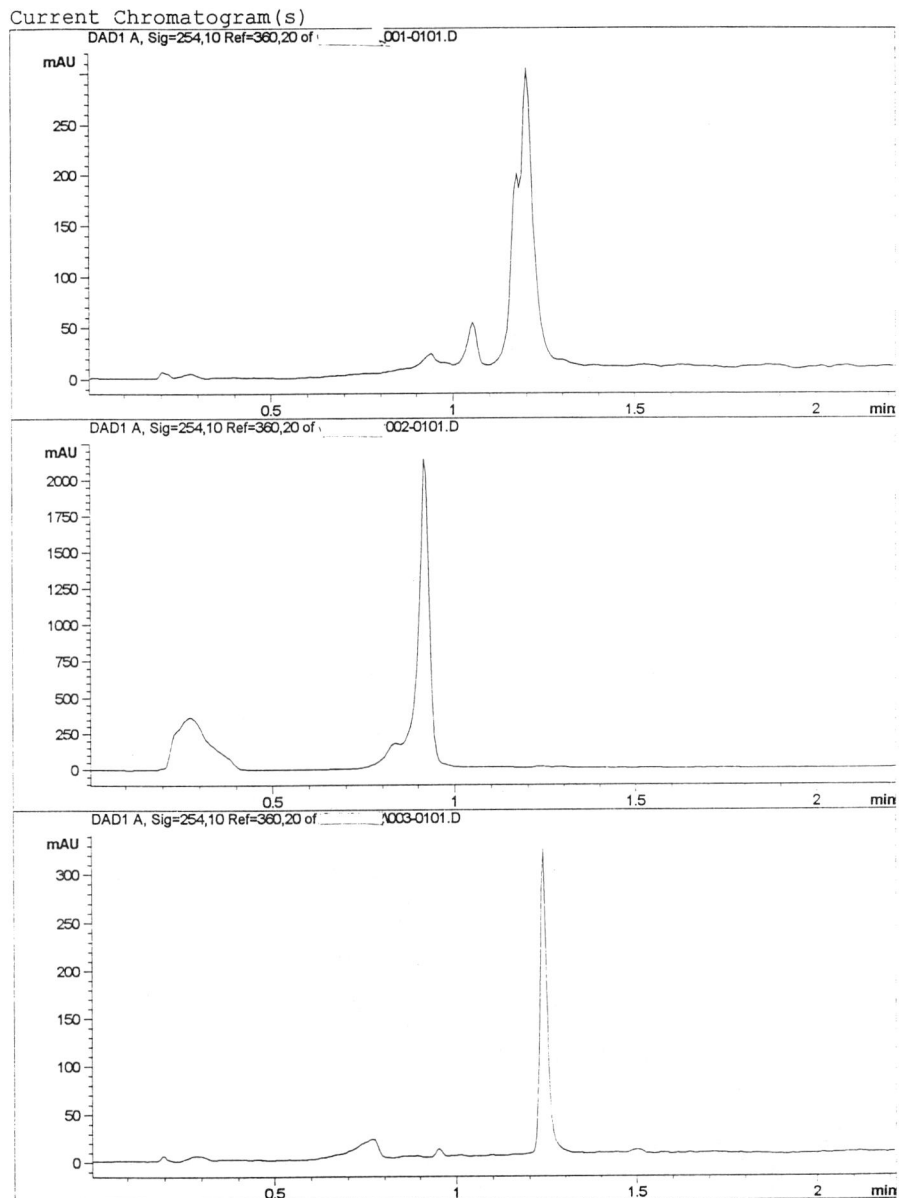

Figure 7. High Speed chromatograms from a 50 x 4.6 mm, 3.5µm column allowing 480 separations/day.

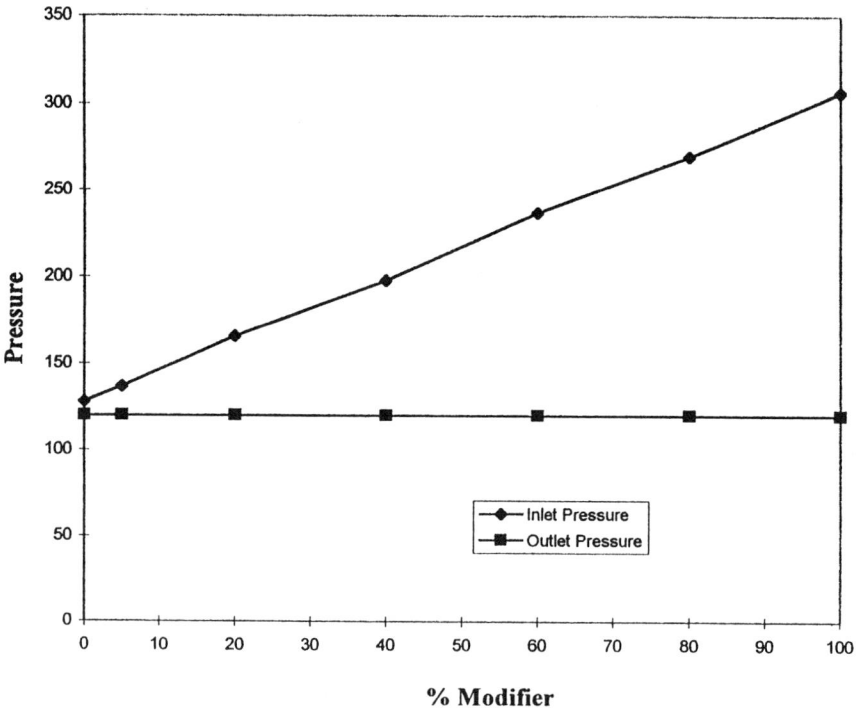

Figure 8. Pressure drop across the 50 x 4.6 mm, 3.5µm column, as a function of modifier concentration.

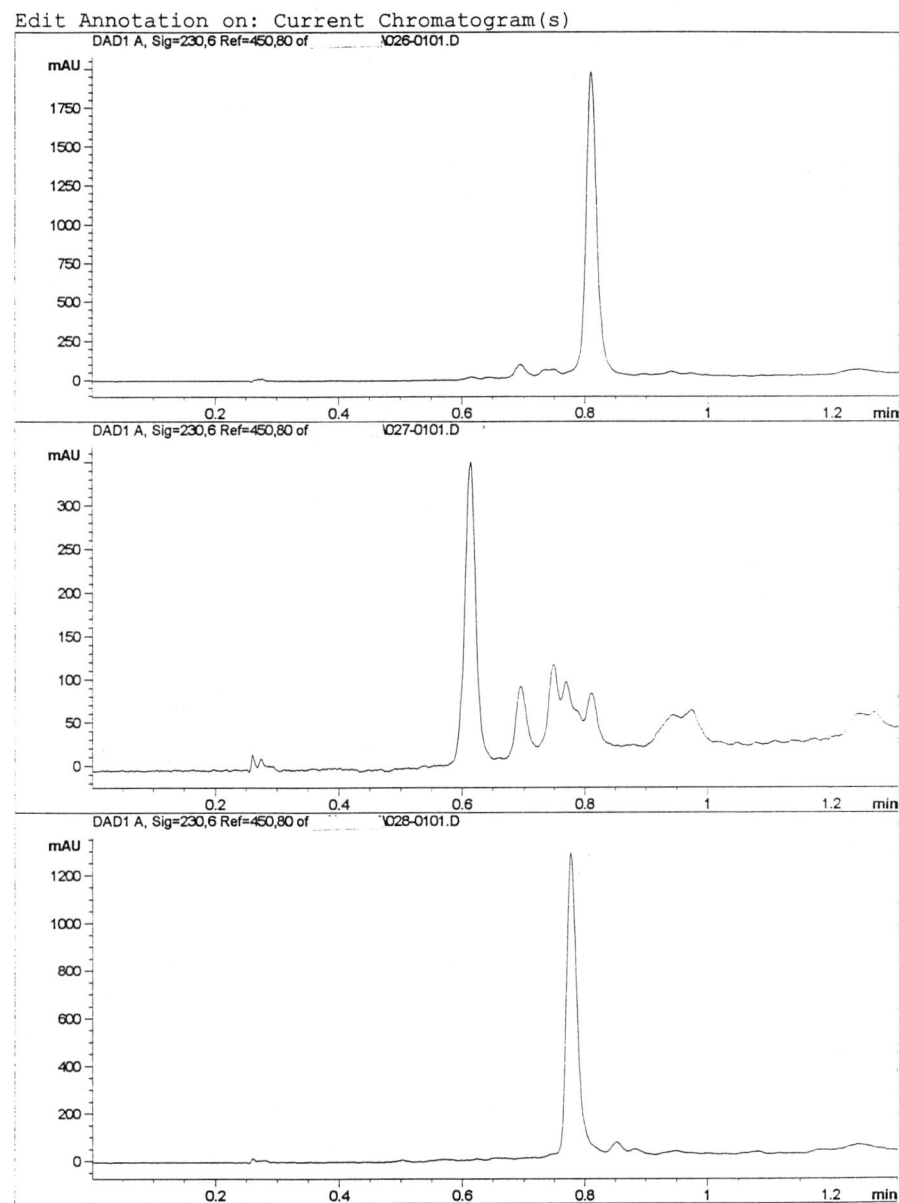

Figure 9. High speed chromatograms from a 125 x 4 mm, 5µm column allowing 770 separations/day.

column, as shown in **Figure 10**. The main component was contaminated by as many as 40 other species, which made up the majority of the sample. The largest injection contained 12.6 mg of the total mixture. Since the compound of interest represented approximately 30% of the total absorbance, this corresponds to nearly 4 mg of the compound of interest on a 4 mm column.

As expected, larger volumes of polar sample solvent distorted and broadened the peaks. The late eluting contaminants were shifted toward the earlier eluting peak of interest. Note that the larger peaks are distorted by the fact that the detector is overloaded. The noisy appearance of the top of the peak indicates no light is passing through the flow cell at that wavelength. Under such conditions, the detector is probably highly non-linear.

The peaks retained adequate integrity to allow peak collection. A fraction was collected between the minimum at approximately 3.2 min. and the second minimum at 4.7 min. Despite the broadening, the peak of interest could be collected and reinjected for a purity check, as shown in **Figure 11**. In this example several small peaks were co-collected with the peak of interest.

Concentrated Injections. . In another experiment, extremely concentrated samples of a few representative compounds were prepared and relatively small volumes injected. The compounds included procaine, imipramine, antipyrine, and caffeine. Each compound appeared to be nearly miscible with methanol, producing solutions up to 0.68g/ml without apparent saturation (except caffeine).

At least 5-10µL, containing as much as 6.8 mg, were placed on the 125 x 4mm CN column and were subsequently recollected. Upon injection into the mobile phase, initial conditions of mostly carbon dioxide, all of the compounds appeared to remain in solution. There was no sudden increase in inlet pressure following an injection. This is a strong indication that these compounds are highly soluble in the mobile phase.

This pumping system (10 mls/min) accommodates 10 mm columns at optimum velocity with a sample capacity for these compounds >40 mg. For significantly higher capacity, pumping systems with much higher flow rates are required. Direct scale-up suggests up to 680 mg in 0.5-1 ml could be loaded on a 41 mm column (with 200ml/min total flow).

Caffeine is much less soluble in methanol than the other compounds and appeared to be saturated at <20mg/ml. As much as 0.55 mg could be injected on-column.

Another relatively dirty experimental compound was dissolved at 440 mg/ml in methanol and 10 µl were injected onto the 4 mm diameter column. Again, there was no indication of precipitation following the injection. A chromatogram, shown in **Figure 12**, shows that several large peaks precede the peak of interest. The main peak was collected between the minima at 6.1 and 7.9 minutes, then reinjected, as shown on the bottom of the figure. Minute amounts of contaminants summing <0.6% are barely discernible on the baseline. Since the peak collected is the last peak to elute, the absence of traces of the earlier eluting peaks in the

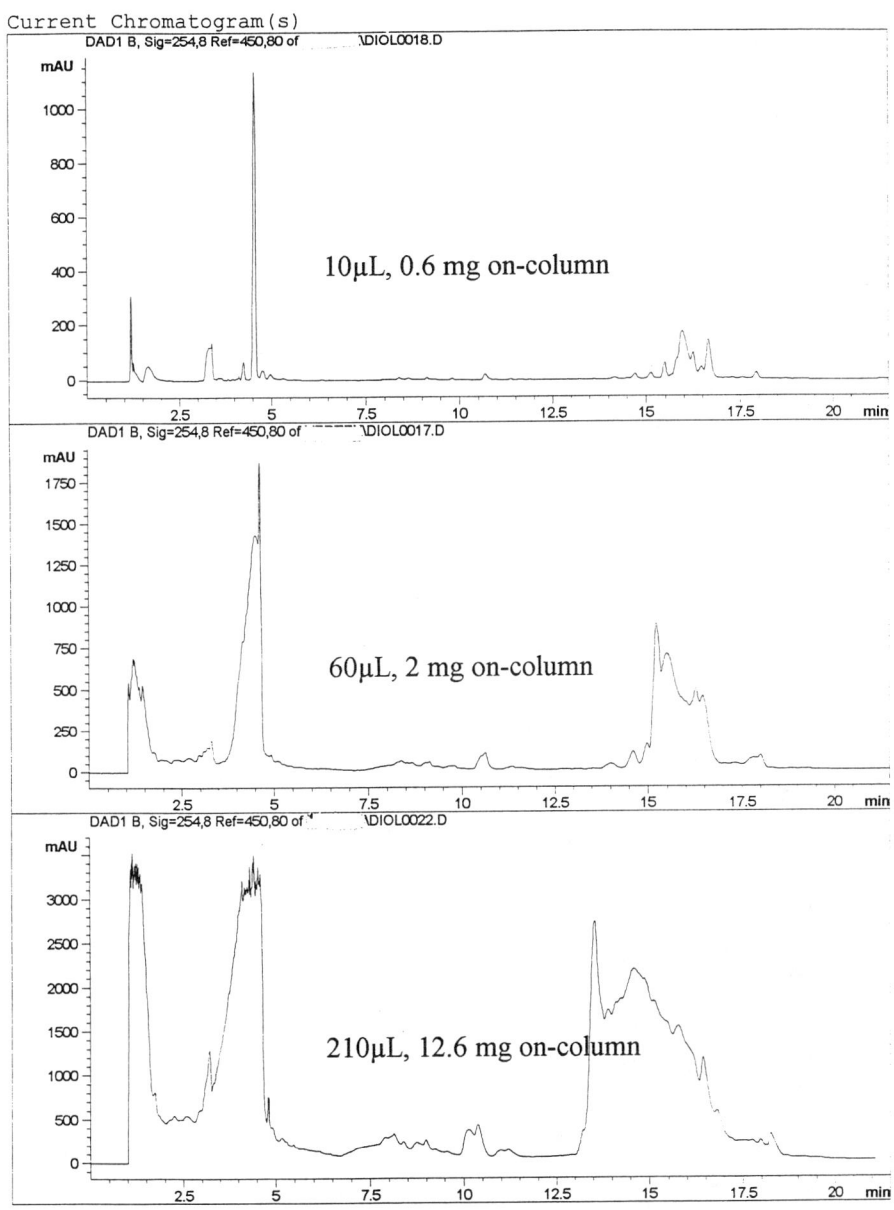

Figure 10. Chromatograms of successively larger volumes of an experimental synthetic mixture, showing the effect of sample solvent volume on peak shapes.

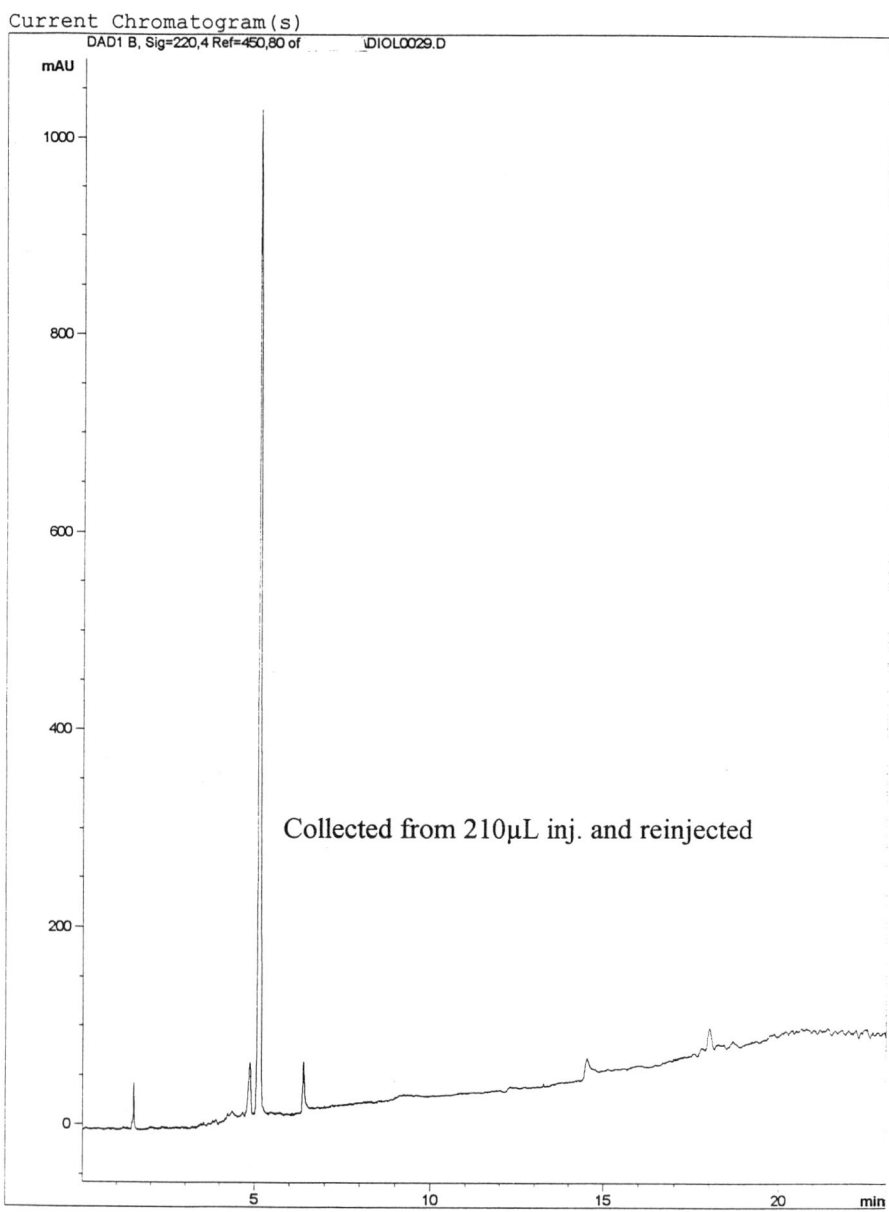

Figure 11. Chromatogram of a collected fraction showing purity.

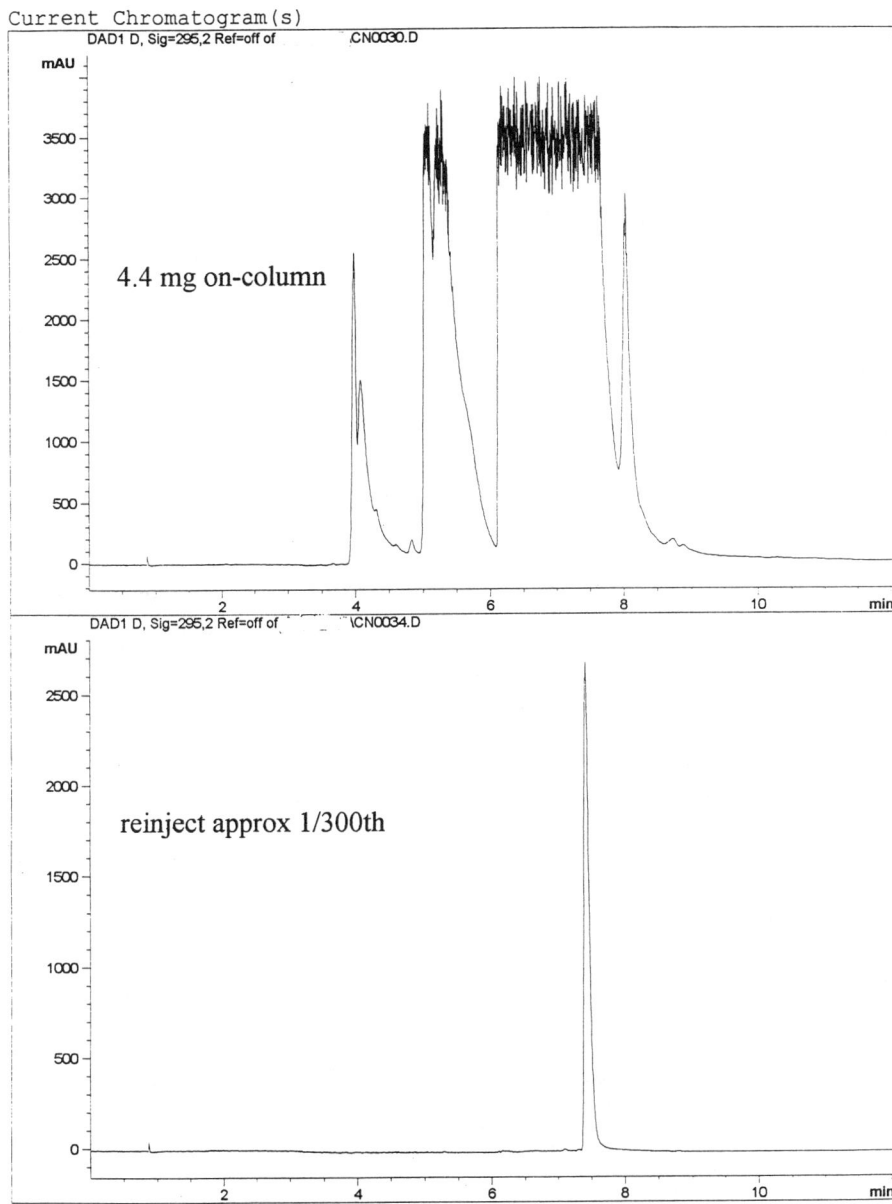

Figure 12. Top. Chromatogram of a 10 μl injection of 4.4 mg of an experimental substance, showing several major contaminants preceding the compound of interest. Bottom. A chromatogram of the effluent collected between 6.1 and 7.9 minutes of the first injection, showing no appreciable carry-over of the preceding peaks into the main peak.

reinjected chromatogram indicate that the collection system does not suffer from peak tailing or carry-over.

Discussion

Considerations in Reducing Cycle Time. Cycle time (time from injection to injection) is dictated by the sum of the gradient delay time, the run time, and the re-equilibration time. If fast enough, the autosampler access time adds nothing to the cycle time.

Gradient Slope. In screening applications, gradients are necessary in order to cover the widest range of solute retentivity in the shortest time possible. For very high throughput, it is desirable to use extremely steep gradients.

A good "rule of thumb" for determining the maximum slope of a gradient is to change conditions such that retention changes by no more than a factor of 2 during each column holdup time. Changing conditions any faster results in part of the column length not contributing to the separation of easily eluted components (they elute at t_o).

Another "rule of thumb" in packed column SFC states that doubling the modifier concentration halves retention. Thus, doubling from 5 to 10%, 10 to 20%, 20 to 40%, etc. each requires at least one column holdup time. On the other hand, equilibration between the stationary and mobile phases requires some finite time, usually 3 to 5 holdup times, suggesting that somewhat slower gradients should yield more reproducible results.

Clearly, a shorter column and/or a higher flow rate decreases the column holdup time and increases the allowed slope of the gradient. However, there are pumping/mixing considerations that degrade performance with very steep gradients, which are discussed later.

Gradient Delay Time. High pressure mixing of binary fluids requires a device with a finite volume. Perturbations in composition vs. time are mixed or averaged out by this volume. At a given flow rate, the time required for a change in composition at the pumps to reach the head of the column is called the gradient delay time. Mixers are intended to act as band broadening devices; to average out sharp, undesirable variations in composition caused by pump refill strokes. The obvious desire to minimize gradient delay times must be weighed against baseline noise and composition consistency. It should be clear that the mixer design is extremely important in obtaining both smooth composition gradients and short gradient delay times.

The commercial SFC used employs an "RC" (resistive-capacitive) hydraulic network to mix the fluids and dampen pressure pulsations. Such networks have a time constant dictated by the dimensions of the components, which are often set to be greater than the period of either pump. For example, if each pump delivers 100 μL/stroke and one pump is intended to deliver 1% of 1 ml/min, the slower pump would refill only once every 10 minutes! If the mixing network were intended to give perfectly smooth composition, it must average out variations over the entire period of the slowest pump, which would require an internal volume > 10 mls! Such large volumes are unrealistic, since the gradient delay time would also be 10 minutes!

In order to achieve reasonable gradient delay times, typical HPLC pumps, which have fixed stroke lengths, must compromise by specifying baseline ripple at 5 or 10% modifier (not 1%). Increasing the minimum modifier concentration to 5% increases the frequency of the slower pump and decreases the mixer volume required by a factor of 5, to 2 mls. Decreasing the stroke volume at the same flow rate and % modifier also increases the pump frequency. Decreasing the stroke volume decreases the required mixing volume and the gradient delay time.

In SFC it is often important to operate down to 1% modifier, due to the non-linear, enhanced solvent strength **(see Figure 3)** accompanying the first small additions of modifier. The default mixer has a dead volume of 1.7 mls, to accommodate operation at 1% modifier. At 5% modifier, 5mls/min, with a 2 ml mixer, the gradient delay time is 0.4 min or 24 sec., which is too long for truly high speed screening. Such a mixer actually adds 48 wasted seconds to a run time (24 at the beginning and 24 at the end).

The SF pump used has a variable stroke length, and a variable speed. With a 20 µl stroke and a minimum modifier concentration of 5%, the optimum mixing volume decreases to 0.4 mls, and the mixer time, constant (at 5 mls/min), drops to 4.8 sec. A 0.4ml mixer was used in place of the 1.7ml mixer.

Without a shorter stroke, the mixing volume cannot be decreased without increasing noise on the baseline and degrading retention time stability. Mixing tees with almost no dead volume (and, consequently, almost no mobile phase residence time) do nothing to average out composition perturbations caused by pump refill strokes. Larger mixers can average out the oscillations caused by an inadequately compensated pump, but the output will have inaccurate flow and composition.

Re-Equilibration Time. An important aspect of the high diffusivity in carbon dioxide based fluids is the rapid re-equilibration of the column to initial conditions. Rapid re-equilibration is aided by a short gradient delay time, high flow rate, and a short column. To determine the minimum delay time between the end of one run and the start of the next, the gradient delay time should be added to the time required to pump 3 to 5 column volumes. The column hold-up time of a 50 mm long column is only 0.1 min at 5 mls/min. Thus, 3 to 5 column volumes correspond to 0.3 to 0.5 minutes, and the total delay should be on the order of 0.4 to 0.6 minutes.

Autosampler Sample Injection Time. The autosampler injection time should not significantly extend the cycle time. On the other hand, screening requires the use of steep gradients, and the column must re-equilibrate to initial conditions between runs. Injection before the completion of re-equilibration effectively increases the initial modifier concentration seen by the next sample, and negates part of the gradient. Thus, if the autosampler access time is less than the re-equilibration time, a delay should be added to avoid premature injection, since the maximum throughput has been reached. For the 50mm column, operated at 5 mls/min, re-equilibration time takes up to 0.6 minutes. The autosampler has up to 36 seconds to withdraw the next sample, push it through the valve, and switch the valve. Longer columns require longer re-equilibration delays and allow longer autosampler access times.

For very rapid analysis (shorter columns, higher flow rate), the autosampler could be prompted to load the "next" sample into the needle or valve during the "current" run.

Pumping Considerations. There have been many instances where only slightly modified HPLC pumps have been used to pump carbon dioxide. This always results in inaccurate flow and composition. In HPLC, the pumps deliver "incompressible" fluids. Properly designed SFC pumps need to be much more complex than HPLC pumps, since the fluids pumped are highly compressible, and compressibility of the fluids is a strong function of pressure.

Single Piston vs. Dual Piston Pumps. Single piston HPLC pumps employ a rapid refill stroke followed by a short, rapid compression stroke. In HPLC, the backward movement of the piston during refill must create a partial vacuum in order to draw in liquids at atmospheric pressure. Such pumps tend to cause problems when used in SFC. During a rapid refill, carbon dioxide at approximately 65 bar in the supply cylinder is suddenly expanded into a space at a much lower pressure. The rapid expansion can cause substantial adiabatic cooling. As a worse case, carbon dioxide dry ice can form, freezing the pump.

In dual piston pumps, one piston slowly fills while the other delivers flow. There is never a significant pressure drop within the pump, preventing significant adiabatic cooling.

The Inadequacy of Chilling Pump Heads. It has been common for chromatographers to chill the pumphead of HPLC pumps in order to pump liquid carbon dioxide. It has been assumed that, since the chilled fluid was defined as a liquid, it had familiar liquid-like characteristics, including low compressibility. Even though such chilling drops the fluid below its critical temperature and it is defined as liquid, it is still far above its boiling point at atmospheric pressure. The fluid will expand to a gas without an externally applied pressure and is still highly compressible.

The compressibility of carbon dioxide (*12*) is not materially affected by temperature near its critical temperature and remains a strong function of pressure, as shown in **Figure 13**. The compressibility at -20°C, commonly used with external bath chillers, is virtually the same as at 20°C. Chilling the pump head

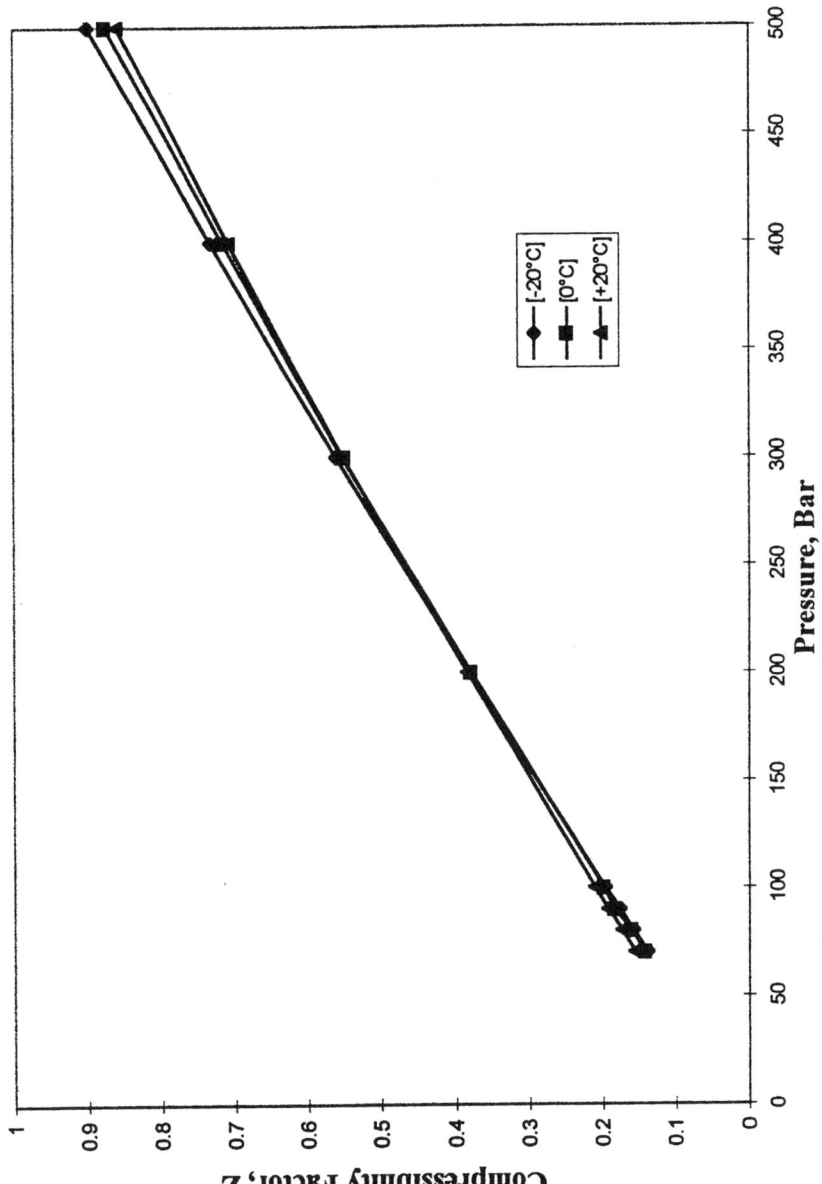

Figure 13. The compressibility of Carbon dioxide as a function of pressure at three temperatures. Chilling the pump head does not make the fluid less compressible.

prevents cavitation (phase separation) but does nothing to minimize the need for extended compressibility compensation. Standard HPLC pumps do not deliver accurate flow or composition of the fluids used in SFC by simply cooling the pumpheads.

Single Compressibility Settings. Most HPLC pumps are built with a hardware specific compressibility compensation (i.e., a specifically shaped cam) set for the compressibility of a normal liquid, usually water. Some more expensive HPLC pumps allow the compressibility compensation for each specific fluid to be set through a keyboard. In most cases, the range of compressibility compensation available is much too small for highly compressible fluids, such as carbon dioxide.

HPLC pumps are not capable of independently, dynamically changing compressibility compensation when the pressure changes. Only pumps designed specifically for SFC have both a wide enough range of compressibility compensation for the fluids of interest and can dynamically change the compressibility compensation based on the current pressure and temperature of the fluid pumped.

Dynamic Compressibility Compensation. In screening applications, a steep modifier concentration gradient is used, which results in dramatic changes in the column head pressure during each run. Since the SF pump must deliver flow to a widely varying head pressure, the carbon dioxide pump must be capable of dynamically changing its compressibility compensation over a wide range. In effect, the pump must calculate what fraction of its total stroke length is required to compress the fluid to each working pressure. Failure to change the compressibility compensation as the pressure changes results in a progressively larger fraction of the pump stroke being wasted as the pressure increases. This produces gross errors in flow and composition, as well as increased baseline noise and decreased retention time reproducibility.

Figure 14 schematically indicates properly compensated pumps deliver accurate flow and composition. **In Figure 15**, part of the stroke of an inadequately compensated carbon dioxide pump is used up compressing the fluid in the pump cylinder, not delivering it to the mixer. While the pump is compressing the fluid, no CO_2 flow emerges. There is no compensation for this loss of flow, and the pump delivers less flow than the set point. The fraction of the expected flow, lost to compression problems, depends on the column inlet pressure. Since the modifier pump is delivering an incompressible fluid, this flow continues. The result is large rapid swings in composition and flow that change with pressure. The pulse dampener/mixer tends to average out such swings, but the averaged results can be grossly inaccurate and noisy.

Future Developments in Screening. The screening experiments reported above do not represent the maximum throughput possible. The 3.5 µm particle column

229

Properly Compensated CO2 Pump

Properly Compensated Carbon Dioxide Pumps deliver flow and composition with accuracy and precision comparible to HPLC

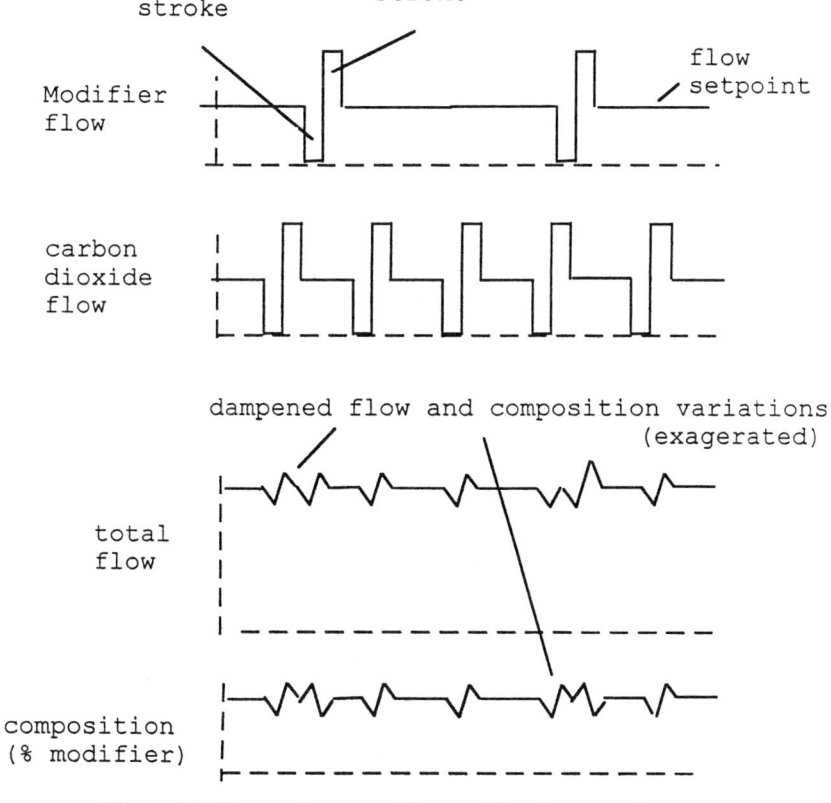

Figure 14. Flow and composition profile of a properly compensated pump.

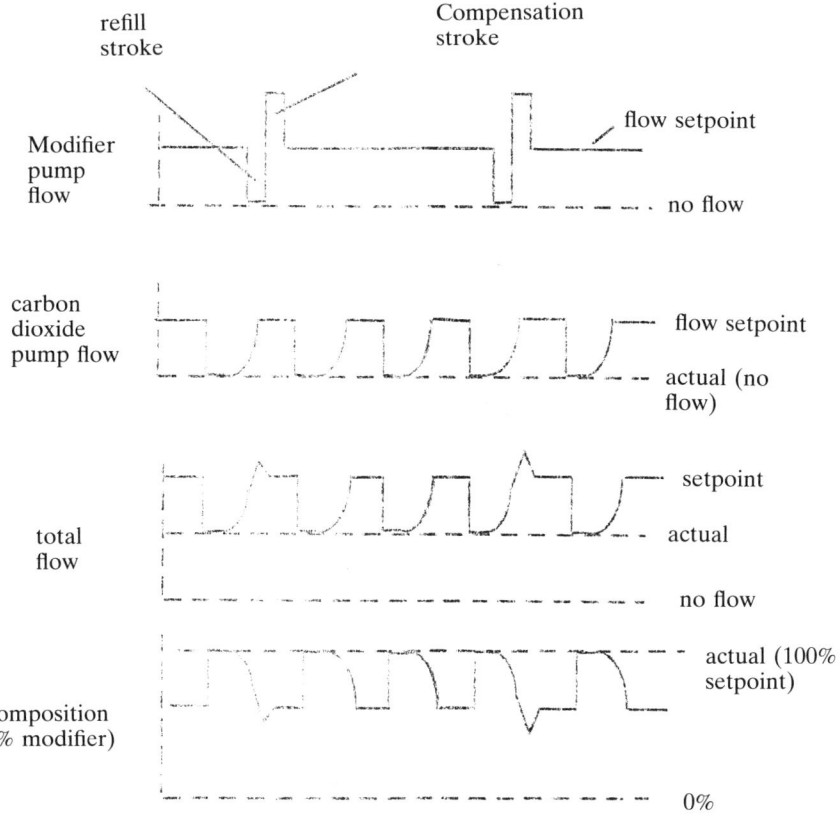

Figure 15. Improperly compensated pumps waste part of the stroke compressing the fluid, not delivering flow.

was operated at only slightly greater than optimum velocity (3 mls/min), the 5 μm column at no more than 2 times optimum (5 mls/min). The pumps are capable of 10 mls/min, but rather than use higher flow rates (with more waste) to achieve higher linear velocities, smaller diameter columns could be used with the same flow rates and gradients.

Smaller particles (i.e., 1.5 μm) exist but not with bonded phases useful in SFC. A few experiments on a 4.6 x 30 mm, 1.5μm experimental diol column generated peeks too fast for a data system to follow. Traditional HPLC chromatographic data systems may be inadequate for high speed chromatography.

Peak capacity decreases with the square root of column length (with the square root of plates) while speed decreases linearly with length. Used carefully, higher information density per time should be possible. However, both shorter columns and smaller particles make extra-column effects more difficult to avoid.

Another important future development is the combination of high speed chromatographic screening with mass spectrometric detection. A number of groups have already reported promising preliminary results using various atmospheric pressure ionization sources with SFC.

Peak Collection Issues

While SFC offers many practical advantages, it tends to require more infrastructure than HPLC. The supply of large amounts of liquid carbon dioxide is nontrivial. A means is required to insure there is liquid carbon dioxide in the pump to prevent cavitation. The carbon dioxide pump must be designed to accurately deliver compressible fluids.

Since a significant amount of the modifier can volatilize on expansion, low modifier concentrations (i.e., <10% modifier) produce no liquid phase to collect. Polar solutes are likely to be trapped inside transfer lines between the outlet of the backpressure regulator (BPR) and the fraction collector. To minimize this problem the low cost liquid pump is often added to the system, with its flow added after the detector but before the BPR. This location requires a high pressure pump capable of pumping against the system outlet pressure. At least one user has employed a low pressure peristaltic pump, with the flow added just downstream of the BPR, with apparently acceptable results.

The expansion of the carbon dioxide tends to generate aerosols in which the solutes can be entrained. Existing interfaces to HPLC fraction collectors may not break such aerosols, creating a potential health hazard since the solutes may escape into the lab air. Separator design is an area requiring further improvement, to simultaneously allow high recovery, low carry-over, and safe operation.

As mentioned previously, up to 10% of the organic modifier vaporizes with the carbon dioxide. If simply vented, some of the "green" advantages of SFC are lost. To prevent such loss, either the fluid must be recycled, or a means must be found to remove the organic liquid from the carbon dioxide stream.

Table I. The stock solutions prepared were:

Procaine	407 mg/ml
Impramine	650 mg/ml
Antipyrine	630 mg/ml
Caffeine	22 mg/ml (satd.)

Table II. Experimental Loadings with Projected Scale-up:

	5 µL	10µL	25µL	projected 40 mm col.
Procaine	2mg	4.1mg	10mg	410 mg
Impramine	3.2	6.5	16.3	650mg
Antipyrine	3.2	6.3	16	630 mg.
Caffeine	0.11	0.22	0.55	22

Conclusions

SFC can provide the chromatographic characteristics of interest in high speed screening, such as high throughput, low pressure drop, less waste, and lower cost. The incentives for developing semi-preparative SFC are: easier, faster solvent removal, lower cost, less waste, and higher throughput. These are all practical advantages for routine use. Carbon dioxide based fluids posses these characteristics irrespective of whether the fluid is defined as supercritical, a subcritical liquid, or a subcritical (dense) gas. The transition from one defined state to another is a non-event, almost always impossible to detect.

Literature Cited

1.) Originally Presented at the 1982 Pittsburgh Conference, Gere, D.R.; Bored, R.; McManigill, D. "The Effect of Column Internal Diameter on Supercritical Fluid Chromatography with Small Particle Diameter Packed Columns" as part of Hewlett Packard Application Note 43-5953-1647, May 1982.
2.) Uyehara, O.A.; Watson, K.M. Nat. Petroleum News, Tech. Section 36, Oct. 4, 1944, 764.
3.) Ashraf-Korassani, M.; Combs, M.T.; Taylor, L.T.; Schweighardt, F.K.; Mathias, P.S. J. Chem. Eng. Data, 1997 42 636-640.
4.) Stahl, E.; Schilz, W.; Schutz, E.; Willing, E. Agnew. Chem. Int. Ed. Engl., 1978 17 731-738.
5.) Berger, T.A.; Deye, J.F. Anal. Chem., 1992 62 1181.
6.) Ashraf-Khorassani, M.; Fessahaie, M.G.; Taylor, L.T.; Berger, T.A.; Deye, J.F. J. High Resolut., 1988 11 352.
7.) Steuer, W.; Schindler, W.; Schill, G.; Erni, F. J. Chromatogr., 447(1988) 287-296.
8.) Berger, T.A.; Deye, J.F.; Ashraf-Khorassani, M.; Taylor, L.T. J. Chromatogr. Sci., 1989 27 105.
9.) Berger, T.A. Packed Column SFC, RCS Chromatography Monographs, Royal Society of Chemistry, Cambridge, 1995, p.99.
10.) Deye, J.F.; Berger, T.A.; Anderson, A.G.; Anal. Chem., 1990 62 615-622.
11.) Berger, T.A. J. High Resolut. Chromatogr., 1991 14 312.
12.) Gas Encyclopedia, Elsevier, Amsterdam, 1976.

INDEXES

Author Index

Andersen, T., 120
Bartle, Keith D., 142
Beck, Thomas L., 67
Berger, T. A., 203
Bowerbank, Christopher R., 179
Bruheim, I., 120
Chester, Thomas L., 1, 6
Clifford, Anthony A., 142
Evans, C. E., 30
Greibrokk, T., 120
Hopkins, Daniel L., 37
Jachwitz, B., 120
Klatte, Steven J., 67
Krouskop, Peter E., 37
Lee, Milton L., 179
Lundanes, E., 120
Martin, Marcus G., 82
McGruffin, Victoria L., 37

Molander, P., 120
Myers, Peter, 142
Olesik, S. V., 168
Parcher, Jon F., 1, 96
Ponton, L. M., 30
Ringo, M. C., 30
Robson, Mark M., 142
Roed, L., 120
Schure, Mark R., 82
Seale, Katherine, 142
Siepmann, J. Ilja, 82
Skuland, I. L., 120
Tao, Yingmei, 96
Tong, Daixin, 142
Trones, R., 120
Wells, Phillip S., 96
Xu, Sihua, 96
Yun, Kwang S., 96

Subject Index

A

Advantages
  Monte Carlo simulations, 37
  packed capillary columns, 143
  packed column supercritical fluid chromatography (SFC), 204
  unified chromatography, 23, 25
Ambient conditions, deceptions, 7–8
Aromatic compounds, separation using combination of GC and SFC by switching mobile phases during same run, 187f

B

Back pressure
  calculation of column, 121–122
  high-performance liquid chromatography (HPLC), 121
Band dispersion control
  comparing for atrazine as function of linear velocity with buffer, 174f
  comparison as function of linear velocity for atrazine, 174f
  enhanced-fluidity liquid chromatography, 171, 173–176
Binary mixtures
  CO_2–methanol Type I system, 13, 14f
  liquid–vapor ($l-v$) behavior, 13, 14f
  methanol–water Type I system, 13, 15
  phase behavior, 12–13
  two-phase $l-v$ region for Type I, 13, 14f
Butane, solvent in liquid chromatography, 8

C

Capacity factor
  predicted pressure-induced changes in solute, for 100 bar increase, 32t
  solute retention, 31
Capillary columns
  classical van Deemter behavior, 144f
  gas and liquid chromatography, 120–121
  minimum reduced plate height versus total porosity for columns with supercritical fluid carrier, dry-packing, and liquid slurry, 145f
  packing, 143–144, 146
  See also Packed capillary columns
Capillary electrochromatography (CEC)
  chromatogram of mixture of polystyrene standards on packed capillary column, 160f
  chromatograms of mixture of phthalates by high-performance liquid chromatography (HPLC) and CEC, 157f
  formation of bubbles, 159
  graph of logarithm of retention factor versus acetonitrile content in mobile phase during CEC, 161f
  packed capillary columns, 133
  retinyl ester profile of liver extracts from polar fox, 135f
  See also Electrochromatography
Carbon dioxide. See Supercritical fluid chromatography (SFC); Universal chromatography
Chemical engineers, solubility of solids in supercritical fluids and liquids in polymers, 2
Chromatographic column, avoiding phase transitions, 12
Chromatographic results, solubility of solids in supercritical fluids and liquids in polymers, 2
Chromatographic separation
  complexity of underlying principles, 82
  phase selections, conditions, and environmental factors, 6
Chromatography
  building unified chromatograph, 22–23
  challenges for researcher, 67
  convection algorithms simulating variety of hydrodynamic conditions, 42
  dependence of resolution on column efficiency, selectivity, and retention factor, 198
  surface interaction algorithms, 45
  See also Stochastic simulation
Coal tar
  separation using two-dimensional capillary SFC–SFC system, 183f
  SFC chromatogram of oil on packed capillary column in unified chromatograph, 154f

237

Column chromatography, blurring traditional boundaries between GC and LC, 179–180
Columns
　back pressure, 121
　capillary and open tubular, 120–121
　longer than conventional liquid chromatography (LC), 25, 26f
　packing capillary columns, 143–144, 146
　relationship between column efficiency per unit time and particle size in packed capillary supercritical fluid chromatography (SFC), 199f
　relationship between resolution per unit time, column length, and particle size in packed capillary SFC, 199f
　robustness, 133, 141
　temperature-efficiency, 122
　universal column efficiency equation, 198, 201
　See also Packed capillary columns
Combinatorial chemistry. See Packed-column supercritical fluid chromatography (SFC)
Compressibility
　carbon dioxide as function of pressure at three temperatures, 227f
　isothermal, 31–32
　predicted pressure-induced changes in solute capacity factor for 100 bar increase, 32t
　relative contributions of mobile-phase, and solute partial volumes to pressure dependence of solute retention, 32t
Computational parameters, simulation input, 40t
Configurational bias Monte Carlo (CBMC)
　deriving force field parameters for non-bonded interactions, 85
　inserting chain molecule atom by atom, 84–85
　reversed phase-liquid chromatography (RPLC), 71–72
　simulation of gas–liquid chromatography (GLC), 84
Convection
　radial flow profiles for laminar and electroosmotic convection, 43f
　three-dimensional stochastic simulation, 42
Cost, unified chromatography, 27
Critical chromatography
　effect of pressure variation on calibration curve for polystyrene standards using CO_2 in THF, 177f
　method for finding critical condition for specific polymer, 175
　separation method using enhanced-fluidity liquid solvents, 175
　studies on use of enhanced-fluidity liquid mixtures as solvents, 175
　telechelic polymer separation, 175–176
　temperature variation to approach critical condition, 175

D

Debye–Huckel theory, assessing limitations of linearization assumption in size exclusion chromatography (SEC) calculations, 68–69
Degrees of freedom, mobile phase, 21
Diffusion, three-dimensional stochastic simulation, 41
Diffusion coefficients
　effect in fluid phase, 46–49, 56
　effect in fluid phase on rate constants, 47t
　effect in surface phase, 56, 59
　effect in surface phase on rate constants, 56t
　relationship of diffusion coefficient of solute and viscosity of solvent, 171
　understanding in high-pressure liquid mixtures, 169, 171
　variation of anthracene in ternary mixtures of ethanol/H_2O/CO_2, 171, 172f
　variation of styrene with temperature for different THF/CO_2 mixtures, 172f
　See also Stochastic simulation

E

Efficiency, universal column efficiency equation, 198, 201
Einstein–Smoluchowski equation, molecular diffusion simulation, 41
Electrochromatography
　capillary electrochromatography (CEC) chromatogram of mixture of polystyrene standards on column with wide-pore packings, 160f
　chromatograms of mixture of phthalates by high-performance liquid chromatography (HPLC) and CEC, 157f
　convection algorithms simulating variety of hydrodynamic conditions, 42
　efficiencies for isocratic CEC on polycyclic aromatic hydrocarbons (PAHs) using HPLC stationary phases, 158t
　electrically driven linear velocity, 155
　electroosmotic flow (EOF) profiles, 155–156
　formation of bubbles during CEC, 159
　packed capillary columns, 152, 155–156, 159
　pressure-driven flow velocity, 152, 155
　surface interaction algorithms, 45
　use of wide-pore silica materials, 159
　See also Capillary electrochromatography (CEC); Stochastic simulation

Electroosmotic flow (EOF)
  electrochromatography, 152
  flow-velocity profiles, 155–156
  See also Electrochromatography
Electrophoresis
  convection algorithms simulating variety of hydrodynamic conditions, 42
  surface interaction algorithms, 45
  See also Stochastic simulation
Electrophoretic migration, three-dimensional stochastic simulation, 44
Elevated temperature liquid chromatography (ETLC), requiring pressure control of mobile phase, 179–180
Eluotropy, principle, 159
Enhanced fluidity liquid
  effect of diffusion coefficients in fluid phase on fluid dynamic behavior of system, 49
  evolution of solute zone profile with varying diffusion coefficients in fluid phase, 52f
Enhanced-fluidity liquid chromatography (EFLC)
  band dispersion control, 171, 173
  chromatogram of di- and monocarboxy terminated polystyrene at critical condition for polystyrene, 177f
  common method of finding critical condition for specific polymer, 175
  comparing band dispersion for atrazine as function of linear velocity with buffer, 174f
  comparison of band dispersion as function of linear velocity for atrazine, 174f
  critical chromatography another separation method using enhanced-fluidity liquid solvents, 175
  difference from subcritical fluid chromatography (SubFC), 19
  effect of pressure variation on calibration curve for polystyrene standards using CO_2 in THF, 177f
  future developments, 176, 178
  instrumental subset, 204
  optimum linear velocity, 173
  pH control for CO_2 added to mobile phase containing water, 173
  requiring pressure control of mobile phase, 179–180
  separating telechelic polymer, 175–176
  small pressure drop across packed columns, 173
  subset of elevated-temperature EFLC, 16, 18f, 19
  temperature variation to approach critical condition, 175
Enhanced-fluidity liquid mixtures
  chromatography, 171, 173–176
  diffusion coefficients, 169, 171
  enhanced-fluidity liquid extraction, 176

future developments, 176, 178
physicochemical properties, 168–171
relationship between diffusion coefficient of solute and viscosity of solvent, 171
solvent strength, 168–169
variation of diffusion coefficients of anthracene in ternary mixtures of ethanol/H_2O/CO_2, 171, 172f
variation of diffusion coefficients of styrene with temperature for different THF/CO_2 mixtures, 172f
variation of solvent strength of methanol/CO_2 mixtures, 169, 170f
variation of solvent strength of methanol/H_2O/CO_2 mixtures as function of added CO_2, 169
Extraction, enhanced-fluidity liquid, 176

F

Flory–Huggins model
  activity coefficient of CO_2 in poly(dimethylsiloxane) (PDMS) as function of volume fraction of PDMS, 109f
  calculation for activity of CO_2 in polymer, 108
  interaction parameter for experimental data, 110f
  relating solubility of compressible fluid in polymer, 108, 111
Flow-through spectroscopic detector, considering for unified chromatograph, 22
Fluid phase
  effect of diffusion coefficient, 46–49, 56
  effect of diffusion coefficient on fluid dynamic behavior of system, 49, 52f, 53f
  effect of diffusion coefficient on rate constants, 47t
  evolution of solute zone profiles with varying diffusion coefficients, 52f
  kinetic behavior, 47, 48f
  radial distribution of solute molecules, 49, 50f
  relationship between total time spent in fluid and surface phases, 49, 54f, 55f
  statistical moments of zone profiles as function of simulation time, 53f
  See also Stochastic simulation

Free energy differences, thermodynamic integration (TI) and free energy perturbation (FEP) methods, 83–84
Free energy perturbation (FEP), method for calculating free energy differences, 83–84

G

Gas chromatography (GC)
  behavior when solute-mobile phase forces are essentially zero, 19, 21

chromatogram of hydrocarbon standards on packed capillary column in unified chromatograph, 148f
compressibility, 33
degrees of freedom, 21
diagram of unified chromatograph, 147f
high pressure GC on packed capillary columns in unified chromatograph, 146, 149
limiting behavior in phase-diagram model, 19
little attention to mobile-phase selection and outlet pressure, 7
plates per second achievable, 150t
pressure as unifying parameter in retention, 33–34
region relative to two-phase liquid-vapor ($l-v$) region, 20f
unified chromatograph, 146
use of temperature control, 7
See also Two-dimensional chromatography
Gas-liquid chromatography (GLC)
alkane partitioning in helium/squalane GLC system, 90–91
combination of Gibbs-ensemble Monte Carlo (GEMC) method and configurational-bias Monte Carlo (CBMC) algorithm, 84
critical temperatures and normal boiling points of linear alkanes as functions of chain length, 87f
determining force field parameters by single-component vapor-liquid coexistence curves (VLCC), 86, 88
exploring fluid phase equilibria governing retention, 83
inside-out approach in design of transferable potentials for phase equilibria (TraPPE) force field, 86
Lennard–Jones potential for van der Waals interactions, 85
multi-component phase equilibria and calculations of Gibbs free energies of transfer, 88, 90
partition constant of solute between phases, 83
partitioning linear and branched alkanes in helium/squalane system, 91
partitioning n-pentane and n-hexane between helium and n-heptane at standard conditions, 88
predicted Kovats retention indices of branched alkanes in helium/squalane system, 92f
predicting accurate critical and boiling temperatures, 86
relative free energy of hydration for conversion of methane to ethane, 88
simulation details and results for n-pentane/n-heptane/helium systems, 90t
simulation methodology, 83–85

single-component phase equilibria and force field development, 85–88
solubility of helium in n-hexadecane as function of temperature and pressure, 92f
testing CBMC/GEMC methodology and TraPPE at elevated temperatures and pressures, 90
thermodynamic integration (TI) and free energy perturbation (FEP) methods, 83–84
TraPPE force field, 85
VLCC for 2,5- and 3,4-dimethylhexane, 89f
VLCC for ethane, n-pentane, and n-octane, 87f
Gibbs ensemble Monte Carlo (GEMC)
reversed phase liquid chromatography (RPLC), 71–72
simulation of gas-liquid chromatography, 84
swapping molecule from one phase to another, 84–85
Gibbs free energy. See Gas-liquid chromatography (GLC)
Glycerides, separation, 127, 128f, 129f
Gradient delay time, reducing cycle time, 224–225
Gradient slope, reducing cycle time, 224

H

Hagen–Poiseuille equation, mean velocity as input parameter, 42
Helmholtz–Smoluchowski equation, maximum velocity as input parameter, 42
High-performance liquid chromatography (HPLC)
chromatograms of mixture of phthalates by HPLC and capillary electrochromatography (CEC), 157f
comparison to SFC for chromatographic purification of samples, 212–213
diagram of unified chromatograph, 147f
difficulty in high throughput screening, 205
efficiencies for isocratic CEC of polycyclic aromatic hydrocarbons (PAHs) using HPLC stationary phases, 158t
high efficiency of nitrated polyaromatic hydrocarbons in unified chromatograph, 151f
limited exploitation of temperature, 6
packed capillary columns in unified chromatograph, 149, 152
unified chromatograph, 146
See also Liquid chromatography (LC)
High speed separations
packed column supercritical fluid chromatography (SFC), 216
plot of inlet and outlet pressure versus % modifier, 218f

series of chromatograms from consecutive wells on microtiter plate, 217f
series of chromatograms using less aggressive column, 219f
High temperature liquid chromatography (HTLC)
  applications, 133
  glycerides separation, 127, 128f, 129f
  high-temperature injector, 138f, 139f
  injection of hydrocarbon wax in cyclohexane, 140f
  See also Packed capillary columns
High throughput screening
  critical point of binary mixtures, 207, 210
  density of carbon dioxide/methanol mixtures as function of pressure and composition, 211f
  difficulties with normal phase high performance liquid chromatography (HPLC), 205
  future developments, 228, 231
  importance of composition when using binary fluids, 207, 208f
  solvent strength as non-linear function of composition, 207, 209f
  van Deemter plots showing relative speed of SFC and HPLC on same columns, 205, 206f
  See also Packed column supercritical fluid chromatography (SFC)
Holy Grail, concept of One World of Chromatography, 1–2
Hydrocarbons
  gas chromatogram on packed capillary column in unified chromatograph, 148f
  separation using carbon dioxide mobile phase transitioning from liquid to gas and then from supercritical fluid to gas along length of column, 196f
  separation using high speed packed capillary supercritical fluid chromatography (SFC), 198, 200f
  See also Polycyclic aromatic hydrocarbons (PAHs)
Hydrocarbon wax
  high-temperature applications, 133
  separation by high-temperature liquid chromatography (HTLC), 140f
Hyperbaric chromatography (HC), lower-than-supercritical-pressure region, 16, 17f
Hypothetical unified chromatography, lattice fluid models, 2

I

Inductively coupled plasma–mass spectrometry
  packed capillary columns, 127, 133
  separation of tetramethyl lead from tetraethyl lead and impurity, 134f

Instrumentation, push to develop, 2
Interfacial resistance to mass transport
  effect on fluid dynamic behavior of system, 61, 63f
  effect on rate constants, 59t, 62f
  kinetic evolution of absorption process with varying, 62f
  statistical moments of solute zone profiles as function of simulation time, 61, 64f
  stochastic simulation, 59, 61
Interphases. *See* Reversed phase liquid chromatography (RPLC)

K

Kamlet–Taft solvatrochromic parameters, evaluating change in polarity of mixtures as function of added liquefied gas, 168–169
Kovats retention indices, simulation results for gas-liquid chromatography (GLC), 91, 92f

L

Lattice fluid models, hypothetical unified chromatography, 2
Liquid chromatography (LC)
  degrees of freedom, 21
  effect of diffusion coefficients in fluid phase on fluid dynamic behavior of system, 49
  evolution of solute zone profile with varying diffusion coefficients in fluid phase, 52f
  evolution of solute zone profile with varying diffusion coefficients in surface phase, 58f
  evolution of solute zone profile with varying interfacial resistance to mass transport, 63f
  good solvent selection, 7–8
  limited exploitation of temperature, 6
  limiting behavior in phase-diagram model, 19
  plates per second achievable, 150t
  pressure as unifying parameter in retention, 34–35
  region relative to two-phase liquid-vapor (l–v) region, 20f
  See also High-performance liquid chromatography (HPLC)

M

Martire and Boehm model
  fitting critical constant-based isotherm to isotherm data for CO_2 in poly(dimethylsiloxane) (PDMS), 116–117
  isotherm equation, 116

lattice fluid model for gas-, liquid-, and supercritical fluid chromatography, 113, 116–117
volume fraction scale, 113, 116
Microchromatography, Raman detection, 162–163, 166
Mobile-phase compressibility, relative contributions, and solute partial molar volumes to pressure dependence of solute retention, 32t
Mobile phase perspective
advantages of unified chromatography, 23, 25
avoiding phase transitions on chromatographic column, 12
binary mixtures, 12–15
building unified chromatograph, 22–23
butane as solvent for liquid chromatography (LC), 8
CO_2-methanol Type I solvent system, 13
combination of downstream pressure control and upstream flow control, 15–16
considering flow-through spectroscopic detector, 22
considering phase behavior of binary mixture, 12–13
continuous one-phase region encompassing ordinary liquid and vapor states, 10, 12
continuum of one-phase fluid behavior, 22, 24f
cost and risk, 27
deceptions of ambient conditions, 7–8
degrees of freedom in mobile phase, 21
enhanced-fluidity liquid chromatography (EFLC), 16, 18f, 19
fitting specific chromatographic techniques into phase diagram, 15–21
fluid continuum available to chromatographers, 11f
gas chromatography (GC) behavior for solute-mobile phase forces essentially zero, 19, 21
good solvent for LC, 7–8
high-pressure loop injector, 22
hyperbaric chromatography (HC), 16, 17f
LC and GC limiting behaviors in phase-diagram model, 19
liquid–vapor (l–v) behavior, 13, 14f
longer columns than conventional LC, 25, 26f
lower-than-supercritical-pressure region, 16, 17f
methanol-water Type I system, 13, 15
optimization in unified chromatography, 25, 27
packed-column unified chromatograph, 24f
phase diagrams describing pure fluids, 8–15
pressure effects on mobile-phase strength, 12
pressure-temperature phase diagram for pure substance, 9f
regions for conventional LC and GC relative to two-phase l–v region, 20f
selectivity tuning by temperature for vitamins A palmitate and D3, 26f
solvating gas chromatography (SGC), 16
subcritical fluid chromatography (SubFC), 16, 18f
supercritical fluid region, 10, 11f
supercritical fluid region of Type I binary mixture, 15, 17f
two-phase l–v region for Type I binary mixture, 13, 14f
unification, 21–23
Models. See Flory–Huggins model; Martire and Boehm model; Sanchez–Lacombe model; Simulations
Molecular level simulations, reversed phase liquid chromatography (RPLC), 68
Molecular parameters, simulation input, 40t
Monte Carlo simulations
advantages, 37
combination of Gibbs-ensemble Monte Carlo (GEMC) method and configurational-bias Monte Carlo (CBMC) algorithm, 84
means to model complex separation systems, 37
See also Gas–liquid chromatography (GLC); Reversed phase liquid chromatography (RPLC); Stochastic simulation

N

Nitrobenzene and dinitrobenzene
high performance liquid chromatography (HPLC) chromatogram of summed integrated intensities of Raman bands during separation, 165f
Raman spectra during HPLC separation, 164f
separation, 163, 166

O

Organometallic catalysts
packed capillary columns, 133
purity check of rhodium catalyst in solvent and as solid by SFC, 136f

P

Packed capillary column chromatography
advantages of packed capillary columns, 143
capillary electrochromatography (CEC) chromatogram of mixture of polystyrene

standards on column with wide-pore packings, 160f
chromatograms of mixture of phthalates by high performance liquid chromatography (HPLC) and CEC, 157f
deuterated solvents as mobile phase for Raman bands, 163
efficiencies for isocratic CEC on polycyclic aromatic hydrocarbons (PAHs) using HPLC stationary phases, 158f
electrochromatography on packed capillary columns, 152, 155–156, 159
eluotropy principle, 159
gas chromatography (GC) of hydrocarbon standards in unified chromatograph, 148f
high pressure GC on packed capillary columns in unified chromatograph, 146, 149
HPLC chromatogram of nitrated polycyclic aromatic hydrocarbons in unified chromatograph, 149, 151f
HPLC chromatogram of summed integrated intensities of Raman bands of nitrobenzene and 1,3-dinitrobenzene, 165f
linear relation between ln k (retention factor) and percentage organic solvent in mobile phase, 159, 161f
packing capillary columns, 143–144, 146
plates per second achievable by different chromatographic techniques, 150t
plot of minimum reduced plate height versus total porosity for capillary columns with supercritical fluid carrier, dry packing, and liquid slurry, 145f
Raman detection as strategy for microchromatography, 162–163, 166
Raman spectra species during HPLC separation of nitrobenzene and dinitrobenzene, 164f
schematic diagram of unified chromatograph for GC, supercritical fluid chromatography (SFC), and HPLC on packed capillary columns, 147f
schematic of micro HPLC chromatograph with microRaman detector, 164f
SFC and HPLC on packed capillary columns in unified chromatograph, 149, 152
SFC chromatogram of coal tar oil in unified chromatograph, 154f
SFC chromatogram of Polywax 1000 polyethylene in unified chromatograph, 153f
shorter wavelength laser light improving sensitivity in Raman scattering, 166
three dimensional two-component phase diagram, 144f
unenhanced Raman detection promising for HPLC on packed capillary columns, 166
unified chromatograph for GC, SFC, and HPLC, 146

van Deemter plots of reduced plate height versus linear mobile phase velocity in HPLC, SFC, and CEC, 144f
Packed capillary columns
advantages, 143
characterization of polymer additive Chimasorb 944 by temperature programming, 131f
column efficiency of all-*trans*-retinyl hexadecanoate and -heptadecanoate as function of column temperature in electrochromatography, 124f
comparing retention and peak shape of polymer additives at constant temperature and temperature program, 126f
compatibility with water as mobile phase, 127
detector compatibility and applications with packed capillaries, 127, 133
effect of temperature on reduced plate height of all-*trans*-retinoic acid, 123f
effect on retention of temperature increase from 50 to 150°C, 127t
electrochromatography, 152, 155–156, 159
flame ionization detection in high performance liquid chromatography (HPLC) in water of *t*-butanol, *sec*-butanol, and *n*-butanol, 132f
high temperature applications, 133, 138f, 139f, 140f
high temperature injector in HTLC (high temperature LC), 138f, 139f
inductively coupled plasma–mass spectrometry, 127, 133
injection of hydrocarbon wax in cyclohexane, 140f
on-column determination of retinoids, 123f
on-line combined solid phase injection–SFE–SFC of methanol/water apple extract containing pesticide fenpyroximate, 133, 137f
purity check of rhodium catalyst introduced in solvent and as solid by packed capillary SFC, 133, 136f
retinyl ester profile of liver extracts from polar fox, 135f
robustness of columns, 133, 141
separation of monoglycerides, diglycerides, and triglycerides by open tubular SFC, packed capillary SFC, and packed capillary HTLC, 128f, 129f
separation of polymer additives by packed capillary column SFC, 130f
separation of tetramethyl lead from tetraethyl lead and impurity, 134f
temperature-column efficiency, 122
temperature-retention in HPLC, 122
use in universal chromatography, 192, 198

viscosity and column back pressure, 121–122
Packed column supercritical fluid chromatography (SFC)
advantages, 204
autosampler sample injection time, 225–226
chromatogram of collected fraction showing purity, 222f
chromatogram of relatively dirty experimental sample, 223f
chromatograms of successively larger volumes of experimental mixture, 221f
chromatographic purification of samples, 210, 212
compressibility of carbon dioxide as function of pressure at three temperatures, 227f
concentrated injections, 220, 224
considerations in reducing cycle time, 224–226
critical point of binary mixtures, 207, 210
density of carbon dioxide/methanol mixtures as function of pressure and composition, 211f
dynamic compressibility compensation, 228
effect of sample solvent volume on peak shapes, 221f
experimental, 212–213
experimental loadings with projected scale-up, 232t
flow and composition profile of improperly compensated pump, 230f
flow and composition profile of properly compensated pump, 229f
future developments in screening, 228, 231
gradient delay time, 224–225
gradient slope, 224
high speed gradient conditions, 213
high speed separations, 216
importance of composition when using binary fluids, 207, 208f
inadequacy of chilling pump heads, 226, 228
large volume injections, 216, 220
peak collection, 216, 220, 224
peak collection issues, 231
peak collection methods, 213
plot of inlet and outlet pressure versus % modifier, 218f
prepared stock solutions, 232t
pumping considerations, 226–228
re-equilibration time, 225
sample solvent capacity, 212
schematic of miniature momentum separator for fraction collection, 215f
schematic of modified SFC for sample collection, 214f
series of chromatograms from consecutive wells on microtiter plate, 217f
series of chromatograms using less aggressive column, 219f
SFC in high throughput screening, 204–210
single compressibility settings, 228
single piston versus dual piston pumps, 226
solvent strength as non-linear function of composition, 207, 209f
unified chromatograph available, 23, 24f
van Deemter plots showing relative speed of SFC and HPLC on same columns, 205, 206f
Partial molar volumes
contributions to pressure dependence of solute retention, 32t
solute retention, 31
Partitioning. See Gas-liquid chromatography (GLC)
Phase diagram model, limiting behaviors of LC and GC, 19
Phase diagrams
continuous one-phase region encompassing ordinary liquid and vapor states, 10, 11f
describing pure fluids in chromatographic mobile phase, 8–15
fitting specific chromatographic techniques, 15–21
illustrating transitions occurring under universal chromatographic conditions, 192, 197f
pressure-temperature for pure substance, 9f
three dimensional two-component, 142, 144f
visualizing effect of pressure and temperature on mobile phase properties throughout length of column, 189, 191f
Phase distribution measurements, theoretical models interpreting retention volume data, 2
Phase transitions, avoiding on chromatographic column, 12
Phthalates, chromatograms of mixture by HPLC and capillary electrochromatography (CEC), 157f
Poisson–Boltzmann methods, size exclusion chromatography (SEC), 68
Polycyclic aromatic hydrocarbons (PAHs)
efficiencies for isocratic capillary electrochromatography (CEC) of PAHs using high performance liquid chromatography (HPLC) stationary phases, 156, 158t
HPLC of nitrated PAHs on packed capillary column in unified chromatograph, 151f
separation at column temperature below and above critical temperature of carbon dioxide mobile phase, 191f
separation of large PAHs using mobile phases transitioning from liquid to gas along length of column, 194f, 195f

separation using comprehensive SFC–GC system, 186f
See also Universal chromatography
Poly(dimethylsiloxane) (PDMS)
 experimental data for solubility of CO_2 in PDMS, 102
 separation using combination of GC and SFC by switching mobile phases during same run, 187f
 thorough study of CO_2 with PDMS, 97–98
 See also Supercritical fluid chromatography (SFC)
Polyethylene, SFC chromatogram on packed capillary column in unified chromatograph, 153f
Polymer additives
 characterization of Chimasorb 944 by temperature programming, 127, 131f
 separation by supercritical fluid chromatography (SFC), 127, 130f
Polystyrene
 capillary electrochromatography (CEC) of mixture of standards on packed capillary column, 159, 160f
 separation of carboxy and dicarboxy-terminated polystyrene standards, 176, 177f
 telechelic polymer separation, 175–176
Pressure
 ambient, 7
 effect on mobile-phase strength, 12
 liquids above normal boiling point, 1
Pressure as parameter in retention
 gas and supercritical fluid chromatography, 33–34
 implications, 35
 liquid chromatography, 34–35
 predicted pressure-induced changes in solute capacity factor for 100 bar increase, 32t
 relative contributions of mobile-phase compressibility and solute partial molar volumes to pressure dependence of solute retention, 32t
 theoretical considerations, 31–33
Pressure control, combination of downstream, and upstream flow control, 15–16
Pressure-temperature path, avoiding discontinuous phase transition, 10, 12
Pumps
 considerations for supercritical fluid chromatography (SFC), 226–228
 dynamic compressibility compensation, 228
 flow and composition profile of improperly compensated pump, 230f
 flow and composition profile of properly compensated pump, 229f
 inadequacy of chilling pump heads, 226, 228
 single compressibility settings, 228
 single versus dual piston, 226

Pure fluids
 phase diagrams describing chromatographic mobile phase, 8–15
 typical phase diagram for pure substance, 9f
Purification
 chromatogram for purity check, 220, 222f
 chromatography of samples, 210, 212

R

Raman detection
 deuterated solvents as mobile phase, 163
 high performance liquid chromatography (HPLC) chromatogram of summed integrated intensities of Raman bands of nitrobenzene and 1,3-dinitrobenzene chromatographic separation, 165f
 improving historical problem of low efficiency of scattering and detection, 162–163
 new strategy for microchromatography, 162–163, 166
 promising detector for HPLC on packed capillary column, 166
 schematic of micro HPLC chromatograph with microRaman detector, 163, 164f
 series of Raman spectra during HPLC separation of nitrobenzene and dinitrobenzene on packed capillary column, 164f
 shorter wavelength laser light improving sensitivity, 166
Retention
 considerations in reducing cycle time for packed column supercritical fluid chromatography (SFC), 224–226
 determining maximum gradient slope, 224
 effect of modifier concentration, 224
 effect of temperature increase, 122, 127t
 methods for predicting, 82–83
 See also Pressure as parameter in retention; Solute retention
Reversed phase liquid chromatography (RPLC)
 adding water/methanol solvent mixtures to modeled stationary phases, 70
 computational methods, 72–73
 consistency of simulation with water/methanol experimental results, 78
 distribution of methanol orientational angles relative to surface normal, 77f
 effect of chain solvation, 75
 effect of hydrogen bonded fluid on stationary phase structure, 73, 75
 effort elucidating underlying driving forces of retention, 68
 examining chain length, surface density, and intermolecular force effects on stationary phases, 69–70

exploring possible onset of phase transitions, 69
modeling retention factor, 67–68
modeling stationary phases in vacuum and in contact with water/methanol mobile phases, 71
molecular dynamics simulations of RPLC interphases, 73
molecular level simulations, 68
partitioning mechanism, 70–71
previous simulations of RPLC interphases, 69–72
relative number densities for chain segments, water, and methanol, 74f
relative orientation of solvent passing from bulk to stationary phase, 75, 78
simulations of alkane stationary phases in vacuum, 71
theories rationalizing observed retention behavior, 68
total occupied volume along z profile, 76f
window potential technique computing potential of mean force for motion of solute into stationary phase, 70
Rhodium catalyst, purity check, 133, 136f
Rice–Whitehead equation, radial velocity profile in cylindrical global frame, 42
Risk, unified chromatography, 27
Rohrschneider-McReynolds scheme, liquid phase characterization, 91

S

Sanchez–Lacombe model
chemical potential for mobile phase of pure CO_2, 111–112
converting solubility data to volume fraction of CO_2 in polymeric phase, 112
determining parameters for CO_2, 112–113
equations-of-state calculations for density of CO_2, 114f
final isotherm equation, 112
interaction parameter as function of volume fraction of CO_2 in poly(dimethylsiloxane) (PDMS), 115f
lattice fluid model for fluid-polymer systems, 111–113
modeling swelling, absorption, and isothermal phase behavior of modifiers, 2
simple isotherm model for CO_2 dissolution in polymer, 111
Screening, high throughput. See Packed column supercritical fluid chromatography (SFC)
Separation strategies, modeling retention as equilibrium process, 67–68
Separation systems. See Stochastic simulation
Separations
high speed, 216

influence of temperature on separation of vitamins, 23, 26f
longer columns than conventional liquid chromatography (LC), 25, 26f
resolution per unit time for fast, 198
See also Enhanced-fluidity liquid mixtures; Universal chromatography
Simulations
reversed phase liquid chromatography (RPLC) interphases, 69–72
See also Gas-liquid chromatography (GLC); Monte Carlo simulations; Reversed phase liquid chromatography (RPLC); Size exclusion chromatography (SEC); Stochastic simulation
Size exclusion chromatography (SEC)
computational methods, 72–73
contour plot of negative salt ion charge density for negatively charged dendrimer inside cylindrical silica pore, 79f
effect of ionic strength on partition coefficient, 67
modeling charged particle as passing from bulk into like-charged cylindrical pore, 73
numerical Poisson–Boltzmann methods, 68
partition coefficient, 72–73
Poisson–Boltzmann equation, 72
preliminary results pertaining to charged dendrimers, 78
Solute retention
capacity factor, 31
contributions to pressure dependence, 32t
function of partial molar volume and bulk compressibility term, 31
See also Pressure as parameter in retention; Retention
Solvating gas chromatography (SGC)
capability of mobile phase pumping system, 181
extension of conventional GC, 16
pressure programming controlling elution, 180
requiring pressure control of mobile phase, 179–180
Solvent, selection for liquid chromatography, 7–8
Solvent strength
enhanced-fluidity liquid mixtures, 168–169
non-linear function of composition in high throughput screening, 207, 209f
variation of methanol/CO_2 mixtures, 169, 170f
variation of methanol/H_2O/CO_2 mixtures as function of added CO_2, 169
Stochastic simulation
advancing each molecule individually through separation system, 38
advantages, 37

comparing supercritical fluid, enhanced fluidity fluid, and liquid as fluid phases, 47
effect of diffusion coefficient in
  fluid phase, 46–49, 56
  fluid phase on rate constants, 47t
  surface phase on fluid dynamic behavior of system, 56, 59
  surface phase, 56, 59
  surface phase on rate constants, 56t
  fluid phase on fluid dynamic behavior of system, 49
effect of interfacial resistance to mass transport, 59, 61
  mass transport on fluid dynamic behavior of system, 61
  mass transport on rate constants, 59t, 62f
effect of reduced diffusion coefficient on rate constants, 57f
equilibrium behavior of system, 46
evolution of solute zone profile with varying
  diffusion coefficients in fluid phase, 52f
  diffusion coefficients in surface phase for liquid chromatography (LC), 58f
  interfacial resistance to mass transport for LC, 63f
hydrodynamic behavior of system, 46
implementing transport algorithms, 38
kinetic behavior of system, 46
kinetic evolution of absorption process with varying
  diffusion coefficients in fluid phase, 48f
  diffusion coefficients in surface phase for LC, 57f
  interfacial resistance to mass transport, 62f
LC case illustrating kinetic and equilibrium information, 47, 49
LC illustrating other hydrodynamic information, 49, 56
mean distance and variance of solute zone profile with varying diffusion coefficients in surface phase, 59, 60f
mean distance and variance of solute zone profile with varying diffusion coefficients in fluid phase, 53f
mean distance and variance of solute zone profile with varying interfacial resistance to mass transport, 61, 64f
means to model complex separation systems, 37
radial solute distribution profiles during kinetic evolution of chromatographic systems, 50f
range in complexity, 37–38
relationship between total time spend in fluid and surfaces phases and distance traveled by individual molecules, 55f
residence time distribution for single sojourn in fluid and surface phases, 51f
residence time distribution for total time spent in fluid and surface phases, 54f
statistical moments of solute zone profiles as function of simulation time, 49, 53f, 59, 60f, 61, 64f
See also Three-dimensional stochastic simulation
Subcritical fluid chromatography (SubFC)
  below critical temperature, 16, 18f
  difference from enhanced-fluidity liquid chromatography (EFLC), 19
  region, 204
Supercritical fluid
  effect of diffusion coefficients in fluid phase on fluid dynamic behavior of system, 49
  evolution of solute zone profile with varying diffusion coefficients in fluid phase, 52f
Supercritical fluid chromatography (SFC)
  ability to use gas chromatography (GC) detectors, 127
  absorption isotherm for CO_2 in poly(dimethylsiloxane) (PDMS), 103f
  accounting for hydrostatic compression of polymer, 98
  activity coefficient of CO_2 in PDMS as function of volume fraction of PDMS, 109f
  apparent gap between gas and liquid chromatography, 96–97
  calculations, 101–102
  capability of mobile phase pumping system, 181
  characteristic parameters for pure CO_2, 113t
  column material, 101
  compressibility, 33
  diagram of unified chromatograph, 147f
  differentiating from GC and high performance liquid chromatography (HPLC), 203
  disparities in measured partial molar volumes of CO_2, 98
  experimental isotherms at 35°C, 104f
  experimental isotherms at 40°C, 105f
  experimental isotherms at 50°C, 106f
  experimental isotherms at 80°C and 100°C, 107f
  experimental values for partial molar volume of CO_2 in PDMS, 100f
  extent of CO_2 dissolution into PDMS, 101
  Flory–Huggins interaction parameter for experimental data, 110f
  Flory–Huggins model, 108, 111
  glycerides separation, 127, 128f, 129f
  influence of temperature on separation of vitamins, 23, 26f
  instrumentation, 101
  interaction parameter as function of volume fraction of CO_2 in PDMS, 115f

literature data for solubility of CO_2 in PDMS at 40°C, 99f
Martire and Boehm model, 113, 116–117
misnaming, 204
models considering mobile phase composition as controlling factor for retention, 97–98
packed capillary columns in unified chromatograph, 149, 152
partial molar volume, 98
plates per second achievable, 150t
pressure as unifying parameter in retention, 33–34
pressure programming controlling elution, 180
relationship between column efficiency per unit time and particle size in packed capillary SFC, 198, 199f
relationship between resolution per unit time, column length, and particle size in packed capillary SFC, 198, 199f
Sanchez and Lacombe model, 111–113
Sanchez–Lacombe equation-of-state calculations for density of CO_2, 114f
schematic of multidimensional capillary SFC–SFC system, 182f
separation of hydrocarbons using high speed packed capillary SFC, 198, 200f
separation of polymer additives, 127, 130f
SFC chromatogram of coal tar on packed capillary column in unified chromatograph, 154f
SFC chromatogram of Polywax 1000 polyethylene on packed capillary column in unified chromatograph, 153f
solubility measurements for pressures up to 100 atm from 35°C to 120°C, 102
theoretical models, 108–117
thorough study of CO_2 with PDMS, 97–98
uncertainty in volumes and compositions of stationary and mobile phases, 97
unified chromatograph, 146
use of back pressure regulator, 204
See also Packed capillary columns; Packed column supercritical fluid chromatography (SFC); Two-dimensional chromatography; Universal chromatography
Supercritical fluid region
no transitions into or out, 10
phase diagram, 10, 11f
Type I binary mixture, 15, 17f
Surface interaction, three-dimensional stochastic simulation, 44–45
Surface phase
effect of diffusion coefficient, 56, 59
effect of diffusion coefficient on fluid dynamic behavior of system, 56, 59
effect of reduced diffusion coefficient on rate constants, 57f
evolution of solute zone profile with varying diffusion coefficients, 58f
kinetic evolution of absorption process with varying diffusion coefficients, 57f
relationship between total time spent in fluid and surface phases, 49, 54f, 55f
statistical moments of solute zone profile as function of simulation time, 59, 60f
See also Stochastic simulation
System parameters, simulation input, 40t

T

Taylor–Aris equation, radial velocity profile in cylindrical global frame, 42
Telechelic polymers
separation at critical condition of polymer backbone, 175–176
separation of carboxy and dicarboxy-terminated polystyrene standards, 176, 177f
Temperature
column efficiency, 122, 124f
effect on reduced plate height of all-*trans*-retinoic acid, 123f
effect on retention, 127t
high temperature applications, 133
retention in HPLC, 122
See also Packed capillary columns
Thermodynamic integration (TI), method for calculating free energy differences, 83–84
Three-dimensional stochastic simulation
absorption and adsorption algorithms for homogeneous and heterogeneous surfaces, 45
convection, 42
diffusion, 41
electrophoretic migration, 44
electrophoretic mobility, 44
first statistical moment or mean distance calculation, 41
input parameters for program, 40t
molar adsorption energy in relation to mean time for desorption, 45
probability distribution, 41
probability of transport between fluid and surface phases, 44–45
radial flow profiles for laminar and electroosmotic convection, 43f
schematic of trajectories of three molecules during four sequential time increments, 39f
second statistical moment or variance, 41
simulation input, 40–41
simulation output, 41
surface interaction, 44–45
verifying accuracy of absorption algorithm, 45
verifying accuracy of diffusion algorithm, 41

verifying accuracy of electrophoretic migration algorithms, 44
See also Stochastic simulation
Two-dimensional chromatography
  combined chromatography, 181, 184
  limitation of typical two-dimensional separations, 184
  progress towards single, combined chromatography form, 3
  schematic of comprehensive supercritical fluid chromatography–gas chromatography (SFC–GC) system, 185f
  separation of polycyclic aromatic hydrocarbons (PAHs) using comprehensive SFC–GC system, 186f
  See also Universal chromatography

U

Unified chromatograph
  elements, 146
  gas chromatography (GC) of hydrocarbon standards on packed capillary column, 148f
  high pressure GC on packed capillary columns, 146, 149
  high performance liquid chromatography (HPLC) of nitrated polycyclic aromatic hydrocarbons (PAHs) on packed capillary column, 151f
  schematic of, for GC, supercritical fluid chromatography (SFC), and HPLC on packed capillary columns, 147f
  SFC and HPLC on packed capillary columns, 149, 152
  SFC chromatogram of coal tar on packed capillary column, 154f
  SFC chromatogram of Polywax 1000 polyethylene on packed capillary column, 153f
Unified chromatography
  advantages, 23, 25
  ambient conditions, 8
  antithesis of concept, 1
  building chromatograph, 22–23
  concept, 1
  cost and risk, 27
  description, 184, 189
  holy grail, 1–2
  idea rather than implementation of concept, 3
  lacking all-encompassing theoretical foundation, 97
  longer columns than conventional LC, 25, 26f
  optimization, 25, 27
  packed column supercritical fluid chromatograph, 23, 24f

push to develop instrumentation, 2
textbook for graduate-level course in separations, 3
unification, 21–22, 24f
unified theory, 2
See also Mobile phase perspective; Universal chromatography
Unified theory, models of equilibrium distribution of solute between stationary and mobile phases, 2
Unified Theory of Chromatography, Martire, 2
Universal chromatography
  capability of mobile phase pumping system, 181
  chromatographic resolution dependence, 198
  combination of supercritical fluid chromatography (SFC) and gas chromatography (GC), 184
  description, 189, 192
  instrumentation, 180–181
  limitations of typical two-dimensional separations, 184
  multiple chromatographic operations, 181, 184
  packed capillary columns, 192
  phase diagram illustrating transitions under universal chromatographic conditions, 197f
  phase diagram with arrows representing mobile phase transitions occurring when column outlet is open to atmosphere, 191f
  plot of relationship between column efficiency per unit time and particle size in packed capillary SFC, 199f
  plot of relationship between resolution per unit time, column length, and particle size in packed capillary SFC, 199f
  plots of reduced plate height versus linear velocity at temperatures above and below critical point of carbon dioxide, 193f
  resolution for fast separations, 198
  schematic of comprehensive SFC–GC system, 185f
  schematic of multidimensional capillary SFC-SFC system, 182f
  schematic of system, 182f
  separation of aldehydes using carbon dioxide at inlet pressure and temperature above critical point and column outlet open to atmosphere, 190f
  separation of aromatic compounds and poly(dimethylsiloxanes) using GC and SFC combination by switching mobile phases during same run, 187f
  separation of dichloromethane solution of household wax using sequential GC–SFC, 188f

separation of hydrocarbons using carbon dioxide mobile phase transitioning from liquid to gas and then from supercritical fluid to gas along length of column, 196f
separation of hydrocarbons using high speed packed capillary SFC, 200f
separation of large polycyclic aromatic hydrocarbons (PAHs) using mobile phases that transition from liquid to gas along length of column, 194f, 195f
separation of PAHs at column temperature below and above critical temperature of carbon dioxide mobile phase, 191f
separation of PAHs using SFC–GC, 186f
separation of standard coal tar using SFC–SFC, 183f
similarity to unified chromatography, 189
transitions occurring in column from liquid to gas, 192
unified chromatography, 184, 189
universal column, 192, 198
universal column efficiency equation, 198, 201
van Deemter equation, 201

V

van Deemter
 classical behavior, 144f
 equation, 201

 plots of reduced plate height versus linear mobile phase velocity in high performance liquid chromatography (HPLC), supercritical fluid chromatography (SFC), and capillary electrochromatography (CEC), 144f
 plots showing relative speed of SFC and HPLC on same columns, 205, 206f
universal column efficiency equation, 198, 201
Viscosity, calculation of liquid, 121–122

W

Wax, household, separation of dichloromethane solution using sequential GC–SFC, 188f